KT-393-333

KING ALFRED'S COLLEGE
WINCHESTER

WITHDRAWN FROM
THE LIBRARY

UNIVERSITY OF
WINCHESTER

KA 0026658 2

Ecology of Soil Organisms

Ecology of Soil Organisms

Alison Leadley Brown, M.A., F.I.Biol.

With drawings by the author

Heinemann Educational Books

Heinemann Educational Books Ltd
LONDON   EDINBURGH   MELBOURNE   AUCKLAND   TORONTO
HONG KONG   SINGAPORE   KUALA LUMPUR   NEW DELHI
NAIROBI   JOHANNESBURG   LUSAKA   IBADAN
KINGSTON

KING ALFRED'S COLLEGE
WINCHESTER

574·5264
BRO   68336

ISBN 0 435 60620 4 (*cased edition*)
       0 435 60621 2 (*limp edition*)
© Alison Leadley Brown 1978
First published 1978

Published by Heinemann Educational Books Ltd
48 Charles Street, London W1X 8AH

Printed in Great Britain by
Cox & Wyman Ltd,
London, Fakenham and Reading

Preface

In the last two or three decades soil ecology has attracted biologists at many levels. From the point of view of research, the field is wide open, offering innumerable possibilities. Population studies of the meso-fauna and their feeding relationships, to mention but one, must receive more attention before the dynamic aspects of soil can be properly interpreted.

At school and university level there are also opportunities for original study. Soils of various types are readily available in both urban and rural situations. This and the fact that a number of books have recently been published on the taxonomy of soil organisms thereby speeding the task of identification, may be the reason why students are turning more and more to the study of soils and the ecology of soil inhabitants.

It is with students in the upper school and in their first year at university in mind, that this book has evolved. It does not attempt to be more than an introduction, although personal inclinations on the part of the author have led to enlargement on some topics and perhaps to greater emphasis being placed on soil fauna rather than its flora. If, however, within the short space available, suggestions have been prompted for projects which can usefully be undertaken, then it will not have failed.

I should like to acknowledge my thanks to Rosaline Murphy for supplying information on *Talitroides dorrieni* and to Dr Roger Lincoln of the British Museum for reading the section concerning *T. dorrieni*. I should also like to express my thanks to Paul Simmonds of the Agricultural Development and Advisory Service for assisting with work on *Tetranychus urticae*.

Contents

Preface   *page* v
Introduction   ix
1. The Environment of the Soil   1
2. Estimating Some of the Chemical and Physical Properties of Soil   14
3. Classifying the Fauna of Soil   21
4. Soil Micro-organisms   35
5. Soil Arachnids and Other Small Animals   44
6. Larger Soil Animals   54
7. Adaptations to Subterranean Life   70
8. Methods of Estimating Populations of Soil Organisms   77
9. The Soil Community at Work   85
10. Dynamic Aspects of the Soil Community   92
11. · Man and the Soil   100
Glossary of terms not explained in the text   109
Bibliography   110
Index   113

Introduction

To all biologists the word 'ecology' must be familiar for whether we set out to investigate an area of woodland, an estuary or the seashore, an ecological approach provides the basis upon which a study of any environment is made.

Soil contains a variety of organisms living together as communities and occupying habitats suited to their requirements of food, physical conditions, and space. Different soils support different populations of animals and plants. These diverse populations and habitats within the soil, together form an ecosystem.

In studying an ecosystem there is an increasing tendency nowadays to make links between the various disciplines. Soil offers research possibilities which are attractive to the chemist, physicist, and biologist. It involves problems which can often only be solved by employing all three subjects, to which we might also add geology. On the other hand, there are many opportunities open to the biologist for making autecological studies or investigations of populations and natural communities occurring in a wide range of soil types.

Sixty or seventy years ago this book would have been much easier to write, Little was known about the microbial life of the soil and the part played by microscopic organisms in the general soil economy. Nowadays geologists have contributed a vast literature on the formation of soil and chemists have made important advances concerning its mineral structure. With the advent of the electron microscope and the use of sophisticated experimental techniques, biologists are now able to offer an explanation to account for the close relationships existing between micro-organisms and the energy cycles operating in the soil. Despite these advances, or perhaps because of them, we are aware that there are wide gaps in our knowledge of the functioning of many of these micro-organisms, especially what part they play in the detritivore chain and their contribution to the process of humification.

Faced with such a wide field of study and the host of micro-flora and fauna, all of which make their individual contribution to the soil community, it is a difficult task to select those which should find a place in a book of this size. Since ecology is the study of organisms in relation to their environment, the stage must be set by considering the soil as a place in which to live. This is the aim of Chapter 1. The organisms themselves and the methods, both behavioural and structural, they adopt in order to make a living in the soil, are the subject of subsequent chapters. Chapters 9 and 10 attempt to outline some of the food relationships which exist among soil animals and the more difficult subject of energy flow between the various feeding levels. The detritivore chain resulting in essential humification processes assumes even greater significance than in other systems. The final chapter attempts to assess the more important aspects of man's interference with the soil ecosystem.

1 The Environment of the Soil

Six hundred miles off the western coast of South America, and lying on the Equator, are the Galapagos Islands. This is an archipelago which, geologically speaking, is of relatively recent origin. All the islands are volcanic and some show signs of very recent eruption. Naked lava covers much of the land around the shores, which are devoid of plant life save for a few encrusting lichens, the *Brachycereus* cactus and opuntias, which have contrived to obtain a foothold and to derive sufficient nourishment from the black lava for growth (Figure 1.1). In such places there is an almost complete lack of soil as we know it. It may be salutory to consider for one moment, what the Earth would be like without soil, the all-provider of food, shelter, and safety from predators for innumerable species of animals and plants.

In this chapter we shall be considering the general characteristics of soil and in what ways it provides an environment in which plants can grow and in which the smaller soil animals and micro-organisms can live and reproduce.

Figure 1.1 Cactus growing on naked lava in the Galapagos Islands. (*Photo by courtesy of David Hosking*)

1

Soil formation

In the early part of the present century emphasis was placed on the chemical breakdown of substances in the soil. The biological processes involved were often neglected probably because little was known about them. Recently we have come to realize that micro-organisms play a vital role in the complex chemical changes taking place in the soil.

To the physicist or chemist 'soil' implies the layer of disintegrated material overlying solid rock, whilst to the biologist, soil is the mineral substrate containing organic material in which plants can take root.

Rain, heat from the Sun, and frost can all cause the splitting of fragments from rocks later to be washed down into the seas, where, in the course of many millions of years, they form layers of sediment so thick that they themselves become compressed into new rocks. Later, through earth movements, these sedimentary rocks may be raised above sea level to become dry land, which in turn is the subject of further denudation (Figure 1.2).

Many parts of the Earth have been twice and three times submerged and elevated, each time to undergo once more the round of disintegration, transport as particles to the seas, compression, and upheaval. Other parts of the land have never been submerged but some mountains, in their long exposure to weathering forces, have become rounded by erosion.

The mineral components form the skeleton of the soil and closely associated with these minerals is the organic matter derived from the vegetation growing in the soil. The agents of decomposition of the parent material to form the mineral components are mainly physico-chemical, while the decomposition of the organic matter is brought about predominantly by soil organisms with physico-chemical agents playing a minor part.

It is evident that in many cases the differences between various soils is due to the different rocks from which they were formed. However, they are not necessarily related to the rocks which they overly, for glaciers in past ages have been responsible for grinding down the rocks over which they pass and transporting this fine material from one area to another. Wind is also a transporter of soil, as witness the dust bowls in the central southern states of America.

In Britain the colonization by plants and the changes in the soils consequent upon changes in vegetation, is a remarkable story. After the glacial periods, the last of which ended about 15 000 years ago, a warmer climate prevailed. As the glaciers melted much of the finer material was washed away in the torrential rivers, leaving stones, gravels, and sand. Tundra-type vegetation prevailed: lichens, mosses, scrub willows, and birch. As conditions became warmer still, pine, hazel, elm, and oak spread from the continent. Later the climate became moister and mixed forests of oak, alder, and ash were widespread. Later still when it became wetter and cooler, peat bogs were formed by the decay of vegetation. The analysis of pollen, retrieved from different depths in the peat, has established this succession.

The dense forests, which covered much of Great Britain until the Middle Ages, had a profound effect on our soils, the most important being the introduction of organic matter which provides a source of food and energy to populations of living soil organisms.

(a)   (b)

Figure 1.2 Sedimentary rock showing (a) thinly bedded metamorphosed shale with irregular quartz veining and (b) folding of bedding planes due to compression, with vertical rift in the rock face

Climate, therefore, has a mechanical effect upon the rocks, fragmenting them into progressively smaller particles, the agents being wind, water, ice, and gravity. At this stage in the development of a soil, it is characterized principally by its original material. Later on this soil will acquire its individual character under the influence of local climate and vegetation. There is a chemical decomposition of the mineral components as well, resulting in the release of soluble minerals which are either absorbed on to the surface of colloidal particles, forming new compounds, or are removed by leaching.

Chemical weathering is, however, a complex process involving a number of reactions – oxidation, reduction, hydrolysis, and carbonation, to mention but a few.

Organic decomposition

In the initial stages of soil formation the vegetation and its associated fauna are poorly developed, starting with colonization by simple fungi and bacteria. The products of organic and inorganic decomposition in these early stages develop together and as they proceed, the soil becomes further fragmented to finer particles, the soil spaces becoming also smaller, thereby increasing the water-holding capacity of the soil. This, combined with an increase in the amount of organic matter and nutrients, means that the soil can support higher plant life such as grasses which assist in binding the soil. The process enables a greater multiplication of soil micro-flora and micro-fauna which further degrade the organic material accumulating at the surface. The end product of this degrading process is the production of *humus* and the pioneers of decomposition are eventually replaced by litter and soil species, often typical of the soil in question. The accumulation of organic matter at the surface increases the depth of soil but, at the same time, the upper organic layers become separated from the influence of the underlying parent material by these transition zones.

The early stages of decomposition of aerial vegetation are assisted by insect defoliators and by herbivorous mammals and may progress quite rapidly at certain times of the year. Once the vegetation falls to the ground, the rate of decomposition increases but will be dependent upon a number of factors such as the nature of the plant material, the humidity, and the surface temperature. The presence of lignin in older vegetation will retard the process, while in a well-aerated soil, in which there are large numbers of earthworms and sapro-phagous organisms, decomposition will be accelerated. However, decom-position by micro-organisms cannot go on indefinitely, although they bring about important changes in the composition of the soil which make it more acceptable to the larger detritus-feeding organisms such as earthworms, insect larvae, mites, collembolans, and so on. They aerate the soil and thereby create conditions favourable for the renewed activity of the micro-species.

The rate of decomposition of plant material is also dependent upon the relative amounts of carbon (C) and nitrogen (N) present. In soil with a high C:N ratio, decomposition is slower because the oxidation of carbon, by bac-teria, fungi, and actinomycetes, inhibits nitrification and the formation of humus. The foliage of softwoods such as alder, ash, and elm, with a low C:N ratio, decomposes more rapidly than the leaves of pine and larch, which have a high C:N ratio. This is apparent even to the casual observer if litter from deciduous woodland is compared with that taken from beneath a pine forest.

The formation of soil profiles and soil types

Collembolans and mites are often present in the very early break-down stages, even those occurring on scant lichens growing on bare rock. With the accumu-lation of soil and successional growth of mosses and grass, the fauna becomes more diverse and insect larvae, worms, and other larger detritivores make their appearance. Breakdown of their faeces is brought about by micro-organisms and so, in a well-aerated soil, the cycle proceeds and humus extends to a greater depth.

The interaction of climatic, chemical, and biotic processes, gradually leads to the formation of horizontal layers, termed *soil horizons*, each having distinct chemical, physical, and biological characteristics. The horizons together form the *soil profile*. Soils which show a well-defined profile are sometimes called 'zonal', 'azonal' soils being those which lack a distinct profile. Theoretically it is easy to make a classification in this way, but there are many intermediate forms of profile which, in practice, are difficult to interpret.

To the biologist some knowledge of the processes involved in soil formation and the reciprocal effects of the organic and mineral components which influence the composition of the soil community, is obviously important. However, digression into an account of innumerable soil types would be impossible here and must be limited to those most significant biologically.

Mull and mor soils

The distribution of the organic material in the soil profile varies with the kind of decomposition process. Forest soils provide excellent examples of organic soils and are of two principal types, named by the Danish pedologist, P. E. Muller, *Mull* and *Mor*. Mor humus tends to be acid and to be associated with more fungal growth and less bacterial activity. Fungal hyphae often bind the surface litter into a matted mass usually distinguishable into three layers: the surface layer comprising fallen leaves and twigs which are not decomposed, below which is a layer where decomposition has begun, and below this again is the true humus layer. Mor profiles are commonly associated with conifers, where there is sandy soil with a low calcium content favouring the production of an acid humus, or on heathland.

In a mull soil, plant debris such as fallen leaves, bud scales, and so on, are loosely scattered on the surface. Decomposition is relatively rapid and the products are quickly incorporated in the surface layers. Mull humus develops under trees such as elm, alder, and poplar in a soil with a higher calcium content which makes it either neutral or slightly alkaline. The distinct layers found in mor humus are not present because the calcium compounds encourage a high population of earthworms. The worms, by their burrowing activities, mix the organic and mineral constituents, distributing the humus more deeply so that the organic-mineral zone merges gradually with the underlying parent material.

In cold, damp upland regions, plant remains decay slowly and the organic material accumulates as peat. Such acid moorland soils contain few earthworms to eat or bury vegetable matter.

In porous sandy soils derived from coarse-grained acid rocks, such as granite, soil acidity can be very high. Heavy rainfall causes the soluble minerals, for instance salts of sodium and potassium, to be washed downwards. As this leaching process continues, calcium, iron, and aluminium are also removed from the upper layers and deposited well below the surface.

The development of an acid soil gives rise to podsolized soils or *podzols*. The natural cover changes from being mainly deciduous to a heath and coniferous type. Micro-fungi thriving at the expense of the bacteria, are less able to bring about the degradation of the surface litter which accumulates as a thick layer of raw humus. Figure 1.3(a) shows a typical podzol profile in which the upper layers have been denuded of minerals leaving a pale, leached horizon. The minerals are deposited lower down. As the profile ages, and this may take many hundreds of years, humus washed down from the surface, collects to form a blackish zone of enrichment in the subsoil. Iron and aluminium oxides accumulate in the lower subsoils. This denudation causes a division of the topsoil into an upper humus-containing zone and a leached layer beneath which there is almost no humus.

In a heavy soil, containing a considerable amount of clay, drainage is impeded and leaching restricted. In some forest soils as much as 80 per cent of the mineral matter is absorbed by plant roots and returned to the soil in plant debris. Such soils favour the activities of earthworms and other detritivores which bring about rapid humification, incorporating the organic matter throughout the upper layers. Figure 1.3(b) shows the profile of a brown forest soil in which disintegration of litter results in a narrow litter horizon merging quickly into the humus horizon, leaving only a narrow fermentation layer.

Figure 1.3 Diagrammatic representation of two soil profiles: (a) peaty podzol and (b) brown forest soil with mull. The broken line indicates the upper level of mineral particles

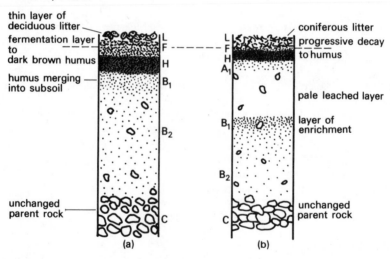

Key

$\begin{aligned}
\text{L} &= \text{Litter layer (undecomposed)} \\
\text{F} &= \text{Fermentation layer (partially decomposed)} \\
\text{H} &= \text{Humus layer (completely decomposed)}
\end{aligned}$

A_1 Horizon = Upper mineral layer
A_2 Horizon = Lower mineral layer
B_1 Horizon = Upper layer of unrichment
B_2 Horizon = Lower layer of enrichment
   C = Parent rock

Conditions for life in the soil

Many factors make up the total environment of the soil as a place in which organisms can live and reproduce. Some soils are almost devoid of life while others offer conditions in which teeming millions of micro-organisms exist. Some of the factors contributing to the total environment of the soil and which are important to its living components are considered below. In very simple terms, soil can be regarded as a solid matrix interspersed with liquids and gases and it will be convenient to discuss these separately.

Soil texture

The mineral constituents of a soil are particles which vary in size, shape, and in their chemical composition. The three main groups of soil particles are sand, silt, and clay and these, in turn, are often subdivided. Definition of size of particle allocated to the different categories is obviously arbitrary but the size fractions most often used are those which were adopted by the International Society of Soil Science in 1927 (Table 1.1):

Table 1.1

Fraction	Particle diameter (mm)
Coarse sand	2.0–0.2
Fine sand	0.2–0.02
Silt	0.02–0.002
Clay	>0.002

A sandy soil is deemed to be one in which sand forms more than 85 per cent, more than 50 per cent in a silty soil, and more than 40 per cent in a clay soil. Loam is a mixture of sand, silt, and clay in various proportions.

Under natural conditions soil particles often occur as aggregations building up into crumbs of different sizes. The crumb structure of a soil is important not only for its living inhabitants but also in agriculture where operations such as ploughing and hoeing are employed, especially on heavy soil, in an endeavour to break down the soil to finer tilth. Winter ploughing is done in order to expose the soil furrows to the alternate effects of freezing and thawing which promote the aggregation of particles. This can be compared to soil cultivation in the tropics where exposure to the Sun has similar effects.

Clay particles greatly influence the physical and chemical properties of soil and are therefore extremely important components. Clay minerals are crystalline in structure and are built up in several plate-like layers, water being able to move the plates apart thus exposing a larger surface. In general, clay particles carry negatively charged ions on their surfaces which attract cations, particularly those of hydrogen, potassium, sodium, and magnesium. The particles become aggregated by the electrostatic bonding of water molecules and metallic ions.

Nucleic amino acids are often found associated with decaying leaves, fungal mycelia, and roots and it is possible that these acids are responsible for the combination of organic and mineral particles to form soil crumbs but the actual mechanism by which this is achieved is still not fully understood. What is certain is that different clay–humus complexes are formed, depending on the type of clay present. Also, the larger detritus-feeders such as slugs and earthworms, during the passage of soil through the gut, do bring about a greater stability between its organic and mineral components.

Pore space

Soil particles do not fill the whole of the available space but leave pores. The particles are irregular in shape, unequal in size, and bunched together. The greater the size of the bunched particles or crumbs the greater will be the percentage pore space, whereas a collection of particles of various sizes will reduce the percentage pore space and the irregular shape of the particles can either increase or decrease this. A clay soil usually has a high percentage pore space which, when wet, will contain very little air; whereas in a sandy soil, dried out almost to *wilting point* (the point at which the force required by plant roots to withdraw water is less than that by which the water is held within its soil pores), nearly all the pore space will be filled with air.

The relative volumes of soil particles, air, and water are clearly important to the organisms living in the soil. In a good pasture soil they are respectively 50, 10, and 40, but these figures will obviously vary according to the soil temperature, rainfall, and the seasonal amount of plant cover.

Soil moisture

Water is perhaps the most important factor at work in the soil, assisting both in its creation and in its destruction. The physical and chemical properties of soil are all influenced by the amount of water present, while from the biological viewpoint, water has a direct effect upon the bacteria, fungi, and other micro-organisms not to mention the larger soil inhabitants.

The amount of water that a soil contains varies within a wide limit ranging from complete waterlogging, when every soil space is filled with water, to complete hydration, when the spaces are filled with air. In between these extremes the pore spaces can be occupied by water to varying degrees and also, within limits, can move from more saturated to less saturated areas.

Water, falling as rain on the soil, penetrates the soil spaces by gravity and, if the rain is heavy and continuous, a stage is reached when the soil can hold no more water and the land becomes flooded. At first the water will drain freely under the influence of gravity, until it reaches an impervious layer. The rate of loss by gravitation will gradually slow down until no more is lost, although the soil may still appear to be moist. The depth of the impervious layer varies both with different soils and at different times of the year. In the case of a heavy clay soil this layer can be very near the surface.

At the point where no more water will drain away, the soil is said to be at *field capacity*. Again, the moisture content of soil at field capacity differs and will vary with changes of temperature, with depth, and with the height of soil above the water table. It also differs from one soil to another. For a sandy soil 20 to 35 per cent is an approximately correct figure, for a clay soil 25 to 35 per cent, and 50 per cent or more for an organic loam. The importance of this first stage lies in the fact that since all air in the soil spaces is now replaced by water, the respiration of plant roots is affected, there being a significant reduction in the rate of gaseous diffusion leading to a depletion of oxygen and a build-up of carbon dioxide.

The second stage of water loss from a soil takes place by evaporation from the surface, by absorption of water by plant roots, and by the activities of soil fauna. This process will continue until wilting point is reached. It is worth mentioning that before this point is reached, plants will show signs of wilting (that is at a pressure of about 10 atmospheres) and will recover if more water becomes available. On the other hand, should this not be so, permanent wilting sets in (at about 30 atmospheres) from which the plant will not recover. This second stage, therefore, ranges over the available water in the soil which becomes depleted by evaporation, by plants, and, to a certain extent, by soil animals. The amount of this available water reserve is critical and varies from one soil to another and from one area of the soil to another because of the movement of water from wet to dry zones.

The soil may still hold some water in the finest pore spaces or absorbed on the soil colloids but the third stage involving the removal of this residual water, can only be brought about by artificial drying.

Considered in another way, that is from the point of view, of distribution of moisture, there are four types of soil water.

(a) *Gravitational water* is that which drains by gravity through the soil, transporting valuable minerals in solution.

(b) *Capillary water* is that held in the pore spaces and plays a similar part to gravitational water, both being readily available to soil organisms.

(c) *Osmotic water* is that held in close contact to clay particles and is less available.

(d) *Hygroscopic water* forms a very thin film round the soil particles and is strongly absorbed by them, very little being available to soil organisms.

The quantitative estimate of soil moisture is complicated by the fact that measurements made by drying do not give a true idea of the amount of moisture available to plants and animals. Measurement of the suction force with which water is held by the soil gives a better indication. This is expressed as the soil pF, defined as the logarithm of the suction force, in centimetres of water, with which the soil water is in equilibrium. This logarithmic conversion gives a pF scale with values ranging from 0 to 7. The actual measurement of pF is difficult although it does reflect the energy relationship between soil and the moisture it contains. These relationships are summed up in Table 1.2.

Table 1.2 Relationship between the various states of water in soil and the suction pressure needed to remove them. (*From Gray and Williams (1971)*)

	Suction pressure needed to remove all water measured as		
State of water	cm water suction	pF	Atmospheres
Gravitational	0–300	0–2.5	0–0.3
Capillary	300–15 000	2.5–4.2*	0.3–15
Osmotic	15 000–150 000	4.2–5.2	15–150
Hygroscopic	>150 000	>5.2	>150

* Permanent wilting point of higher plants.

Different soils can be subjected to various suction pressures (pF) and after equilibrium at each pressure has been reached, the percentage moisture content can be determined. By plotting pF against moisture content a soil moisture characteristic curve is arrived at. Figure 1.4 shows typical curves for clay, loam, and sandy soils. The biological importance to micro-organisms of the relationship of soil moisture to pF is evident. For each of the soils the energy required by a micro-organism or plant root to obtain water is different. For instance at a soil moisture content of 10 per cent a micro-organism will have to exert a suction pressure in excess of pF 1.5 in sand, 4.0 in loam, and 5.2 in clay in order to obtain water.

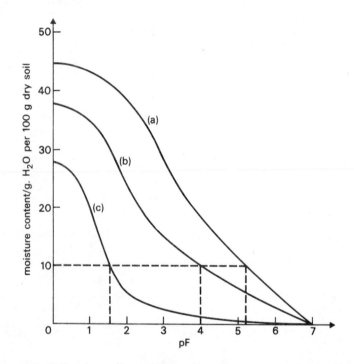

Figure 1.4   Soil moisture characteristic curves. Drying boundary curve of (a) a clay soil, (b) a loam soil, and (c) a sandy soil. (*After Griffin (1963) and modified by Gray and Williams (1971).*)

For soil organisms, hygroscopic water is of little importance since it forms such a thin layer, no more than 20 molecules in thickness, around the soil particles and because it is held by a tension of many atmospheres, it is virtually unavailable. Capillary water, however, is of great importance for it contains nutrients in solution and ensures that the soil atmosphere is at or near saturation. The *relative humidity* of the soil atmosphere is expressed as the percentage of the amount of moisture needed to saturate the air at the same temperature and pressure. Since an increase in temperature will increase the capacity of the atmosphere to hold water vapour, the relative humidity will vary with temperature. But the drying power of the air can vary with temperature even though the relative humidity remains constant. It is this evaporative power of the air at a given temperature, usually called its *saturation deficit* (the difference between the actual water vapour pressure and the maximum water vapour pressure) which is important to cryptozoic organisms for many micro-arthropods cannot survive in conditions of low humidity since their loss of moisture to the surrounding atmosphere is closely related to its evaporative power.

Soil atmosphere

From the point of view of the soil itself the relative proportions of air and water in the soil spaces largely determine the characteristics of the soil but the air channels are also the living spaces for plant roots and the micro-fauna of the soil. The effects of the moisture within the pores have been discussed in the previous section but where a soil is completely waterlogged, anaerobic conditions must prevail. Except under these circumstances, the larger soil spaces will be filled with air, the oxygen and carbon dioxide content of which is all important to the organisms living there.

Organic debris or compacted soil at the surface prevent the free flow of air between soil and atmosphere. This may lower the concentration of oxygen in the soil air while that of carbon dioxide will build up due to the metabolic activities of the soil organisms. In cultivated loam at or near optimum moisture content, the percentage by volume of oxygen in soil air can be 20.00 compared to 21.00 in atmospheric air and of carbon dioxide, 0.5 compared to 0.03 in air. There may also be other gases such as ammonia and methane, the result of microbial activity, and in the areas around plant roots or where there are

Figure 1.5 Podzol profile showing iron pan. (*Photo by courtesy of Peter Beale*)

aggregations of bacteria and fungi, the concentrations of gases will be altered.

An exchange of gases between the soil and the atmosphere above takes place through those pore spaces which are filled with air. This exchange will tend to reduce the concentration of carbon dioxide in the upper layers of the soil but has a decreasing effect in the lower regions. Diffusion of gases within the pore spaces, from areas of high to lower concentrations takes place all the time and this movement will be accelerated if the pore spaces are not filled with water since gaseous diffusion takes place more slowly through water. Thus the rate of soil aeration is directly influenced by moisture.

Changes in the gaseous content from one area to another affect the distribution of micro-organisms. Indeed, the balance between aerobic and anaerobic conditions can be a very fine one, anaerobic micro-environments occurring in most soils which are condusive to the growth of anaerobic bacteria including the important nitrogen-fixing organism, *Clostridium pasteurianum.* Anaerobic conditions may often lead to the accumulation of sulphides and ferrous iron and to the proliferation of bacteria capable of reducing these substances. The reddish-brown layers of an iron pan in heathland soils is witness to their activities (Figure 1.5).

Concentrations of carbon dioxide in the soil atmosphere can influence the pH providing a source of carbon for autotrophic bacteria and at the same time limit fungal activity.

Temperature

The temperature of a soil is related to the amount of solar radiation falling on the surface but there are a number of factors which determine the amount of energy which is absorbed such as the degree of slope and the angle at which the Sun's rays strike the surface. The amount of heat required to bring about a change of temperature will depend on the nature of the soil surface, its texture, and moisture content. In the northern hemisphere, light-coloured soils absorb less heat than darker ones and it must also be remembered that where there is a lot of vegetational ground cover, much of the radiated heat will be intercepted before it reaches the soil. This is most noticeable in forests with a close tree canopy, especially in coniferous woodland where, in addition, there is a thick layer of litter insulating the soil beneath.

In organic soils, heat radiation at the surface is not so quickly transferred to the lower layers as in soils with a lower organic content. The temperature of the deeper humus and mineral layers may still be cooling down during the early part of the day while that of the air and surface layers is rising. When the temperature of the air and surface layer begins to fall in the latter part of the day, the temperature of the deeper layers continues to rise as night approaches, although less sharply. In general, the variations in temperature decrease in amplitude with increasing depth and even the surface layer shows less variation than the air above.

Soil temperature cannot be described in isolation from other soil factors. There is, for instance, a close connection between the amount of moisture in a soil and its heat-absorbing capacity, a moist soil usually showing less fluctuation of temperature than a dry one. Nevertheless, a wet soil will conduct heat downwards more efficiently than a dry soil so that although the temperature of the latter at the surface may be higher than that of a wet soil, a few centimetres below the surface their temperatures may be the same. Evaporation of water from the surface dries the surface layers and causes water to move upwards from the deeper layers, but at the same time there is a downward conduction of heat from the surface. As the temperature of the lower layers rises, the rate of diffusion of gases from the pore spaces increases and a complex interaction of temperature, moisture, and aeration results.

The effect of temperature gradients on soil micro-organisms is either to increase or decrease their metabolic processes according to their optimum growth temperatures. Those living near the surface may be subjected to considerable fluctuations during the course of a day, while those living in the deeper layers probably experience only small changes of temperature throughout the year. It is known that cryptostigmatic mites, perform regular vertical movements in the soil, during the 24-hour cycle, which permit them to remain at or near their optimum temperature requirements despite diurnal temperature variations. There is, however, much still to be discovered about the effect of temperature upon the distribution and activities of most soil organisms.

Summary

Although soil composition, moisture, temperature, and atmosphere can all be measured on a macro-scale, the existence of micro-climates within soil is of the greatest importance in considering the distribution of soil organisms. However, the ability to measure such micro-environments is often limited by the absence of sufficiently sensitive techniques.

Because of the constant fluctuation in the various components of the soil environment and the effects of one factor upon another, it is difficult, if not impossible, to relate the behaviour of soil organisms to any single factor. Soil, therefore, offers a very specialized environment and at the same time highly variable conditions to which the animals must adapt. The different ways in which they do so, and the influence of one upon another, will be dealt with in later chapters.

2

Estimating Some of the Chemical and Physical Properties of Soil

There may be many reasons for studying soil which involve tests of its consistency, chemical reactions, and so on. For the ecologist tests will be selected in order to find out why certain organisms live in a particular environment and in what way the nature of the soil is a contributory factor in their distribution. Some of the tests which are more important for an ecological survey of soil are given below.

The area under investigation should be defined on a grid so that the position of soil samples taken from different places within the area can be marked on the grid.

Making a soil profile

A very good general idea of the structure and layering of the soil can be obtained by making a soil profile. This is especially useful if two or more areas are to be compared.

A trench is dug, 1 or 2 metres in length, if possible down to the parent rock material. In moorland soils this may be near the surface but will be deeper down in well-cultivated land. In digging the trench, one side should be kept vertical so that measurements can be made of the depths of the different soil horizons.

In a well-developed profile of a highly leached soil or podzol the horizons can be clearly seen (see Figures 1.3 and 1.5). In immature soils or in dry situations, some of the zones may be absent or poorly developed, but even so the main horizons can usually be distinguished.

Soil profiles can afford useful information about the structure and nature of the soil and of the degree of leaching, while the positions and depths of the various zones exposed in the profile may offer valuable clues as to the distribution of soil organisms.

Taking soil samples

From the previous section and Chapter 1 we have seen that vertical zonation of the soil layers can play an important part in determining the composition of the soil. It is important, therefore, that any device used for taking soil samples should be able accurately to extract a sample from a known depth. One of the instruments that can be used to do this is called a *soil corer*. The corer illustrated in Figure 2.1 consists of a metal tube, 6 cm in diameter and 30 cm long, marked in centimetre graduations. The lower end has a sharp cutting edge. A metal rod, acting as a handle, passes through two holes at the top of the tube. A small refinement can be added to one side of the metal handle in the shape of a round piece of wood, slightly narrower in diameter than the metal tube. This can be used as a ram rod to expel the soil core after a sample has been taken.

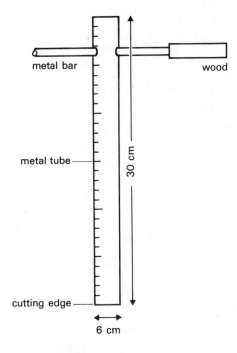

Figure 2.1   Metal soil corer

The corer is thrust into the soil and turned by means of the handle. The depth to which the corer is pushed into the soil will depend on which soil layer is to be sampled. The soil cores are expelled into separate tins, labelled with the depth and position at which each was taken. Samples should be taken from at least six different sites within the area and if a general sample is required, they should be thoroughly mixed; otherwise they should be kept separate. For sampling surface layers an ordinary tin, marked in centimetre graduations may suffice. For sampling very hard soil to greater depths it may be necessary to use a *soil auger*. This is like a very large corkscrew which is screwed into the ground and then withdrawn vertically. Soil samples taken by whatever method can now be subjected to various tests.

From the point of view of the requirements of living organisms in the soil some of the tests described in various books and papers are immaterial. For the purposes of this account, only the factors most affecting the living requirements and distribution of soil organisms have been selected.

Testing soil reaction

Soils are naturally so well buffered that considerable amounts of water can be added to a sample without altering the pH.

Special pH indicators, together with colour charts are available commercially. (British Drug Houses supply an indicator, marketed as the Universal Indicator, together with a colour chart, which will test the pH between the range of 4.0 and 11.0.) A small amount of the soil sample is shaken up with distilled water and allowed to settle. A drop of the indicator is then added to the clear fluid decanted from the sample. The colour produced will give a rough indication of the pH of the sample when matched against the colour chart. For more accurate measurements, a capillator (such as the BDH Capillator set) can be used. A suspension of the sample is made as described above and a drop is drawn up to a fixed mark in a capillary tube and discharged into a small watch glass. The rougher method using the indicator will have established the approximate pH so that the most suitable indicator from the capillator set can now be selected. A drop of this indicator is added to the drop of soil suspension in the watch glass, mixed, and the mixture drawn back into the capillary tube. A comparison of colour can be made against the series of standard colours in capillaries supplied with the set. In this way the pH of the sample to an accuracy of 0.1 units can be made.

Estimating the water content

The absolute water content of soil influences the activity and distribution of soil animals but clearly this will fluctuate according to the rainfall. However, a series of samples collected at the same time in different places (allowing a period of about 24 hours to elapse after heavy rain) will provide reliable comparisons to be made.

Samples, each weighing 10 g are then dried in an oven to about 105 °C, cooled in a desiccator, and reweighed. They are then replaced in the oven and the process repeated until no further loss in weight occurs. The loss in weight, expressed as a percentage of the oven-dried soil, represents the moisture derived from the hygroscopic water of the sample plus some of the capillary water.

Estimating the organic content

An accurate determination of the organic matter in soil is difficult to obtain, but measurement of the oxidizable humus will give a reasonable estimate of the organic content.

5 g of oven-dried soil which has been stored in a desiccator, is placed in a crucible and heated to red heat for 30 minutes. It is then cooled in a desiccator and reheated, the process being repeated until there is no further loss in weight. The loss in weight, expressed as a percentage, represents approximately the amount of oxidizable organic matter present. Although this method offers a reliable means of comparing different samples, it must be noted that it does not measure all the organic material present, particularly the carbonates. A correction for this can be made by adding a small amount of ammonium carbonate solution to the cooled sample and reheating to 105 °C in an oven. The sample is then cooled and reweighed. The gain in weight will represent the amount of carbon dioxide lost by the carbonates during the initial heating.

Estimating soil minerals

The inorganic salts contained in the soil water play an essential part in the growth of plants and therefore, indirectly, in the occurrence and distribution of the animals associated with them.

In this context, the minerals of greatest importance to plants are compounds of nitrogen (usually present as nitrate (V) (nitrate), nitrate (III) (nitrite), and ammonia compounds), and phosphorus. These are usually estimated by colorimetric methods.

Before such an analysis of soil can be made, a soil extract must be prepared which is then subjected to various tests. These, in turn, require the preparation of various reagents to act as colour indicators. Soil indicators are now produced commercially for those engaged in agriculture, and even gardeners can employ colour-testing as a rough guide for estimating mineral deficiences in their soils. For more sophisticated methods, various textbooks of agricultural chemistry should be consulted (see Bibliography).

Soil atmosphere

Little is known about the chemistry of the air below soil level, although we do know that it differs from that of atmospheric air and that its composition fluctuates, consequently influencing the distribution and activity of soil animals.

Measurement of the oxygen and carbon dioxide content of the soil air have been made by Harley and Brierley (1953) using an ingenious piece of apparatus consisting of lengths of plastic (PVC) tubing sufficiently large to provide an adequate sample of soil air for analysis. The basic tube (Figure 2.2) is about 30 cm long and 3 cm in diameter. It is sealed at one end and fitted at the other with a short length of narrow PVC tubing and a well-greased screw clip. A length of strong wire, formed at one end into a hook, is soldered on to the screw clip. The tube is tested for leaks before use by closing the screw clip and submerging it in water.

A tube, fitted in this way, is buried in the soil with the wire hook projecting above ground. In doing so, the soil is inevitably disturbed and must be allowed to settle for a day or two, during which time the screw clip is closed by twisting the wire hook. The tube is then opened and diffusion of gases both into and out of the tube takes place until an equilibrium is reached, which will take about a

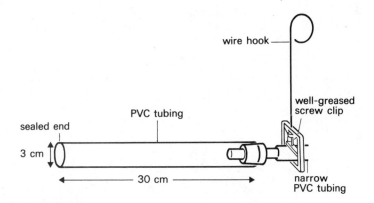

wire hook

well-greased
screw clip

PVC tubing

sealed end

3 cm

30 cm

narrow
PVC tubing

Figure 2.2 Apparatus for sampling the soil atmosphere. (*After Harley and Brierley (1953).*)

week. The screw clip is then closed and the tube removed from the soil. The contents of the tube are then analysed by means of a Barcroft-Haldane gas analysis apparatus (see Peters and Van Slyke, 1956). (A number of the basic tubes can be joined together to increase the size of the soil air sample, if this proves necessary.)

Measuring soil temperature

Measurements of soil temperature can be made by means of a *thermistor*. *This* is a resistance thermometer consisting of a temperature sensitive glass bead, less than one mm³ in volume, mounted at the end of an evacuated tube. The resistance of the bead varies greatly with the temperature and in the opposite sense from that of metal. The evacuated tube contains connecting leads joined to the bead by very fine platinum wires so that little heat is transferred from the bead to the connecting leads (Figure 2.3). These are connected to a simple Wheatstone Bridge circuit (Figure 2.4) in conjunction with an ammeter and small dry battery, the whole apparatus being calibrated with a mercury thermometer in a water bath. The apparatus is easy to construct (Figure 2.5) and is designed to measure temperatures in many kinds of micro-habitat.

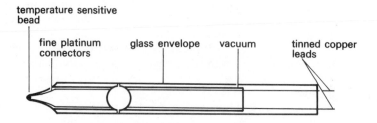

Figure 2.3 A thermistor in section. Long leads are possible because the resistance of the element is high and power demands low so that an amplifier is not necessary. (*After Macfadyen (1957).*)

Figure 2.4 Circuit diagram for thermistor with three temperature ranges and three alternative thermistors. (*After Cloudsley-Thompson (1960).*)

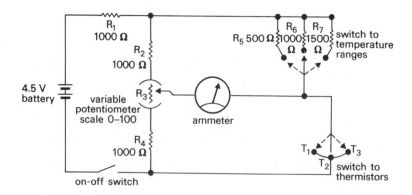

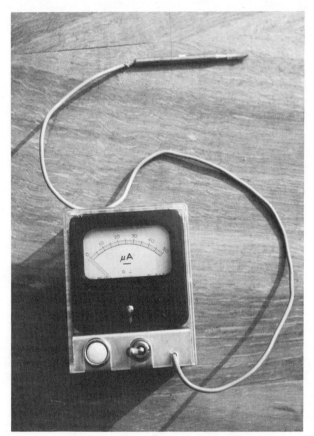

Figure 2.5 Thermistor and ammeter

The graphs in Figure 2.6 were constructed from temperature measurements recorded in a sheltered garden during a March day, made by means of thermistors of the type just described. Two sites were selected for the recordings, one on a mown grass lawn in a position which received maximum sunlight and the other on naked soil in the shade. There was a ground frost in the morning followed by a thaw and sunny weather until dark. 5, 10, and 20 cm lengths of aluminium tubing, closed at the bottom ends, were sunk at each of the sites and temperatures recorded at those depths. The upper end of each tube was corked and the thermistor cable passed through the cork so that the tip of the thermistor recorded the temperature at the bottom of the tube.

From these results it is clear that (a) there was a time lag at each site between temperature changes at the surface and changes in the lower layers, showing the temperature at 10 and 20 cm depth to be still falling at noon. (b) The amplitude of the temperature changes in the lower soil layers was considerably less. (c) The temperature at 20 cm depth in naked soil was at all times higher than that at 10 cm depth. This was possibly due to the higher moisture content of the soil at this site.

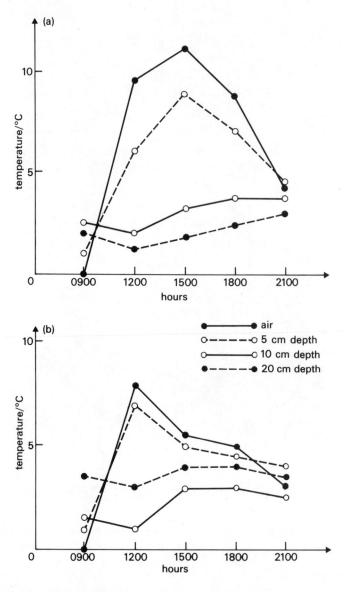

Figure 2.6 Diurnal temperature curves in air and at different depths beneath the surface of (a) grass lawn and (b) naked soil, 17 March, 1975

3 Classifying the Fauna of Soil

Most authorities consider that soil micro-organisms are microscopic animals and plants which are less than 200 μm in size. A convenient, although artificial, method of classifying the rest of the soil fauna is also by using size as a criterion. Figure 3.1 shows an arbitrary arrangement of the various groups classified in this way.

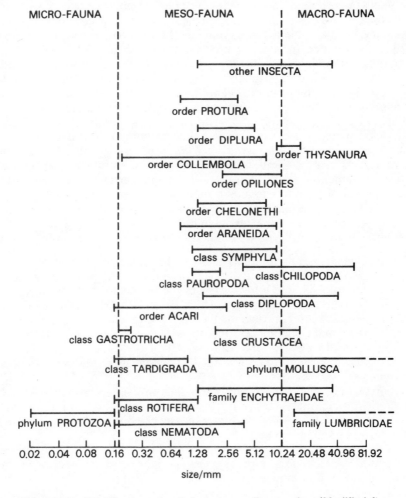

MICRO-FAUNA | MESO-FAUNA | MACRO-FAUNA

other INSECTA

order PROTURA

order DIPLURA

order THYSANURA

order COLLEMBOLA

order OPILIONES

order CHELONETHI

order ARANEIDA

class SYMPHYLA

class PAUROPODA | class CHILOPODA

class DIPLOPODA

order ACARI

class GASTROTRICHA | class CRUSTACEA

class TARDIGRADA | phylum MOLLUSCA

family ENCHYTRAEIDAE

class ROTIFERA

phylum PROTOZOA

class NEMATODA | family LUMBRICIDAE

0.02 0.04 0.08 0.16 0.32 0.64 1.28 2.56 5.12 10.24 20.48 40.96 81.92

size/mm

Figure 3.1 Classification of soil fauna according to size. *(Modified from Wallwork (1970).)*

21

There are other and probably more acceptable methods of grouping the soil fauna. For instance a distinction can readily be made between temporary and permanent inhabitants of the soil. Many insects are examples of temporary soil inhabitants either undergoing part of their development in the soil to emerge as adults or entering the soil to seek temporary shelter as adults when conditions become unfavourable. Larvae of the insect orders Diptera, Coleoptera, and Lepidoptera are temporary inhabitants and we shall describe their activities in greater detail in later chapters. Among the permanent inhabitants are hibernating beetles, thrips, and bugs. All can be called *geophiles*, some like the insect larvae, being active and others inactive geophiles. The insect larvae are well adapted by their feeding habits, movement, and body shape to life in the soil. All the same they must be regarded as temporary inhabitants for most leave the soil as adults, becoming adapted in various ways to an aerial existence.

Animals belonging to the permanent soil fauna, sometimes called *geobionts*, spend their entire life in the soil. Representatives of this category are found among the Protozoa, Nematoda, Annelida, Myriapoda, Acari, Mollusca, and some of the wingless insects including the Collembola.

Another method of classification is based on the distribution of fauna throughout the soil or the inhabitants of each layer of soil in the vertical stratification: those living in the deeper mineral layers, those belonging to the organic layers, and those present in the surface litter. A further group is distinguished as those animals, such as rotifers, tardigrades, and some nematodes, living within the film of water surrounding soil particles or the water-filled soil spaces. However, this form of grouping rather breaks down when one considers the numbers of organisms which exhibit seasonal or diurnal vertical movements within the soil and for some, vertical movements which take them out of the soil altogether and into the aerial vegetation.

A more natural method of classification based on different methods of feeding cuts right across the taxonomic groups and for this reason can be usefully adopted as an alternative method. We can define the following groups:

1. *Microphytic feeders* or those animals feeding upon fungal spores and hyphae, lichens, and bacteria. Insects such as some species of ant, fungus gnats, nematodes, protozoa, and a few molluscs can be included in this group.

2. *Saprophytic feeders* are those which derive nourishment from dead or decaying organic matter and include the earthworms, enchytraeids, millipedes, isopods, Acari, and Collembola. Sometimes they are collectively referred to as scavengers or *detritivores* and can be subdivided into those which are faecal-feeders (*coprophages*), wood-feeders (*xylophages*), and carrion-feeders (*necrophages*).

3. *Phytophagous feeders* are those feeding on living plant material either aerial green leaves and stems (molluscs and some insect larvae), roots (many nematode parasites, symphylids, dipteran, and coleopteran larvae), or woody parts (termites and coleopteran larvae).

4. *Carnivores*. In this group belong the true predators such as carabid and staphylinid beetles, some mites, spiders, pseudoscorpions, centipedes, and some nematodes and molluscs.

This method of grouping soil animals really implies a food pyramid starting with green plants – the primary producers – which provide plant material in the form of dead or decaying fruits, leaves, stems, and roots to the soil ecosystem. The primary producers form food for the primary consumers which in the case of the soil community are the saprophages or phytophages. These, in turn, are the food of the carnivores which may be subdivided into secondary consumers and tertiary consumers, the latter being those carnivores which feed directly on other carnivores.

Methods of moving through the soil offer another means of classification, basically into two groups. The first group are those animals which are small enough to be able to utilize the pore spaces in which to move. These are the animals of the soil water film. The second group are the burrowing animals which actively dig their way through the soil. These include the soil lumbricids, the fossorial Orthoptera and Coleoptera, the fossorial Orthoptera and Coleoptera, and even the mole. Some of the larger animals such as a few lumbricids, millipedes, and centipedes, are able to squeeze through existing channels in the soil.

Here, then, are only a few ways in which soil animals can be grouped. They are by no means exhaustive but serve merely to give an idea of the different ways in which we can observe the soil community.

The next chapter will describe some of the groups of micro-organisms, leaving the meso-fauna and the larger soil animals to be considered in subsequent chapters.

Keys to soil animals

The keys given on pages 24–34 are intended to assist in consigning some of the more important soil animals to their broad taxonomic groups.

For the purpose of such an outline classification, it must suffice to divide soil animals into four main groups – worm-like animals (Key1), arthropods (Key 2) which are by far the largest group, and slugs and snails for which no key is given because few species use the soil as a place in which to live permanently. A fourth group are the larvae of many species of insects which inhabit soil for the whole or most of their larval life (Key 3). Their similarity to worms can be misleading, making identification difficult especially in very small forms. This key has been constructed from the more detailed keys in Bechyne (1956) and Chu (1949). The majority of insect species whose larvae consistently form part of the soil community, belong to the beetles and the two-winged flies. The larvae of certain noctuid moths are also soil dwellers. They can be distinguished from beetle or fly larvae by possessing several pairs of false legs resembling tube feet, on the abdomen. In addition, many species of Hymenoptera inhabit soil. Most construct underground nests in which the larvae are reared.

Our knowledge of the distribution, feeding habits and population biology of soil larvae of different kinds is considerable, yet their taxonomy remains relatively undocumented. The purpose here is to distinguish, in very general terms, between the families of the two insect orders and it must be stressed that such a brief key can be misleading, since a specimen to be identified may differ in characteristics from any of those keyed.

Mites are often encountered in extractions made from soil litter. To assist in distinguishing between the four main groups of terrestrial mites, a separate and more detailed key (Key 4), taken from Evans (1955), is given for this order, the Metastigmata, however, have few soil representatives.

Opposite the number 1 in each key are two alternatives. Select the one appropriate to the animal to be identified. This may give the answer directly to the group to which it belongs or may lead to another number. Continue in this way until you reach the answer. For Keys 1, 2, and 3 this will only give the class, order or family to which the animal belongs.

Where a drawing of an organism also occurs in the text, the figure number is given after the title to the figure in the key. Some figures may only show the diagnostic features mentioned in the key.

More detailed keys listed in the references at the end of the book, will have to be consulted for identification down to genus or species.

1   Key to 'worms'

1 Body divided into rings or segments **2**

    Body not segmented **4**

2 Body composed of more than fifteen segments
    (Class Oligochaeta) **3**

    Body composed of less than fifteen segments. Not
    worms but insect larvae (see Key 3)

3 Small, thin worms, usually grey or white ......   **Enchytraeids**
                      (white- or potworms)

    Larger worms, usually pink or brownish ......   **Lumbricids**
                      (earthworms)

White- or potworm, 15–20 mm. **Enchytraeid**   Earthworm. **Lumbricid**

4 Body somewhat pointed and enclosed in a rigid
    cuticle which limits contraction and expansion   Class **Nematoda**
                      (roundworms including
                      eelworms)

    Microscopic; tail divided into two **5**

Eelworm, *Rhabditis* sp., 1 mm. Class
Nematoda

5 Cilia, constantly beating and arranged in a disc (wheel organ) on the head .... Class **Rotifera** (wheel animalcules)

Body flattened and often covered with bristles Class **Gastrotricha** (water bears)

Wheel animalcule, less than 2 mm. Class **Rotifera** Water bear, 200 μm. Class **Gastrotricha**

---

2 Key to soil arthropods

1 Numerous pairs of jointed legs 2
Three or four pairs of jointed legs 7

2 One pair of limbs attached to each body segment; thoracic limbs differing from abdominal limbs (Class Crustacea) 3
Body elongated, most segments bearing one or two pairs of similar limbs 4

3 Body flattened from top to bottom; seven pairs of thoracic legs; abdominal limbs (pleopods) flattened .... Order **Isopoda** (woodlice)

transverse section through thorax

pleopod

Oniscus asellus (Figure 6.3(b)). Order **Isopoda**

Body flattened from side to side; abdominal limbs not flattened .... Order **Amphipoda** (sandhoppers)

Talitroides dorrieni – transverse section through thorax (Figure 6.5). Order **Amphipoda**

4 Body cylindrical; most segments bearing two pairs
of limbs .... Class **Diplopoda**
(millipedes)

Most body segments bearing one pair of limbs **5**

Iulus sp. – transverse section through body
segment (Figure 6.8(a)). Class **Diplopoda**

5 Animals not more than 10 mm long; body cylindrical; not more than twelve pairs of limbs .. **6**

Larger animals; body flattened with prominent
jaws and, when adult, with fifteen or more pairs
of limbs .... Class **Chilopoda**
(centipedes)

Lithobius forficatus – head and first body
segment (Figure 6.7(a)). Class **Chilopoda**

6 Very small animals (1 mm approx.); when adult,
have nine pairs of legs and body segments fewer
in number; antennae three-branched .... Class **Pauropoda**
(pauropods)

Small animals (more than 1 mm); when adult
usually have twelve pairs of legs; antennae
unbranched .... Class **Symphyla**
(symphilids)

Pauropus sp. – head and first body segment
(Figure 6.6(a)). Class **Pauropoda**

Scutigerella sp. – head and first body segment (Figure 6.6(b)). Class **Symphyla**

7 Body divided into head, thorax, and abdomen;
antennae usually present; wings may be present
(Class Insecta) **12**

Head and thorax fused to form cephalothorax;
without antennae; four pairs of legs (some larvae have only three pairs) (Class Arachnida) **8**

8 Minute, often transparent animals with stumpy, unsegmented legs .... Order **Tardigrada** (tardigrades)

Not as above **9**

Macrobiotus sp., 0.4 mm. Lateral view (Figure 5.7(a)). Order **Tardigrada**

9 Small animals with lobster-like claws (pedipalps) Order **Chelonethi** (false scorpions)

Not as above **10**

False scorpion – head and claws (pedipalps) (Figure 5.2). Order **Chelonethi**

10 Body distinctly divided into two parts – a cephalothorax and an abdomen; spinning glands present .... Order **Araneida** (spiders)

Cephalothorax and abdomen completely joined; no spinning glands **11**

(a)

spinnerets

(b)

Spider (a) dorsal view and (b) ventral view of end of abdomen. Order **Araneida**

Harvestman. Order **Opiliones**

11 Legs usually very long and slender; abdomen segmented .... Order **Opiliones** (harvestmen)

Small or microscopic; legs short; abdomen unsegmented .... Order **Acari** (mites) see Key 4

12 Wings present (Sub-class Pterygota) **13**
   Wings absent (Sub-class Apterygota) **19**
13 Two pairs of wings **14**
   One pair of wings, the second pair represented by
     halteres ....   Order **Diptera**
                                               (two-winged flies)

— haltere

Cranefly, *Tipula* sp., body 20 mm. Order
Diptera

(Adults normally occur only temporarily in soil.
For key to soil-inhabiting larvae see Key 3.)

14 Both pairs of wings membranous **15**
   Forewings wholly or partly chitinized (except
     Homoptera) and used to protect the mem-
     branous hind wings **16**
15 Wings bearing scales ....   Order **Lepidoptera**
                                        (moths and butterflies)

   Wings clear (may be absent in certain castes);
     abdomen usually attached to thorax by a dis-
     tinct waist (gaster) ....   Order **Hymenoptera**
                                      (ants, bees, and wasps)

gaster

Large yellow underwing, *Triphaena*
pronuba. A noctuid moth whose larvae are
soil-living cutworms. Order **Lepidoptera**

Wood ant, *Formica rufa*, 10 mm. Order
Hymenoptera

16 Mouthparts adapted for piercing and sucking ..   Order **Hemiptera**
   Forewings not divided into two regions and may   (bugs)
     be membranous (Sub-order Homoptera)

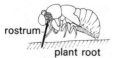

rostrum —
plant root

rostrum —
leaf cells

Cicada nymph (Homoptera). Order **Hemiptera**

Head of plant aphid. Order **Hemiptera**

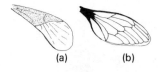

Forewing of (a) heteropteran and (b) homopteran. Order **Hemiptera**

Forewings divided into a leathery base portion and membranous tip (Sub-order Heteroptera) Mouthparts not as above **17**

17 Forewings (tegmina) wholly chitinized, without veins, and meeting down the centre to cover or partly cover abdomen **18**
Forewings leathery with veins; hind limbs much longer than the other two pairs and used for jumping . . . .   Order **Orthoptera** (grasshoppers, crickets, cockroaches)

Grasshopper, *Chorhippus* sp., 20 mm, with one pair of wings extended. Order **Orthoptera**

18 Last abdominal segment bears a pair of curved forceps . . . .   Order **Dermaptera** (earwigs)

Forewings very hard, forming elytra which do not take part in flight; in some, e.g. the staphylinids, the elytra are short and do not completely cover the abdomen . . . .   Order **Coleoptera** (beetles)

(For key to soil-inhabiting coleopteran larvae see Key 3.)

Earwig, *Forficula auricularia*, 15 mm, ♂, with one pair of wings extended. Order **Dermaptera**

Common cockchafer, *Melalontha melalontha*, 25 mm, with one pair of wings extended. Order **Coleoptera**

Burying beetle, *Necrophorus vespilloides*, 18 mm. a staphylinid beetle. Order **Coleoptera**

19 Abdomen with six segments or less; usually possess a forked springing organ and 'ventral' tube   Order **Collembola**
(springtails)

Body more elongated with more than six abdominal segments and no springing organ **20**

springing organ

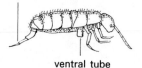

Springtail, 4 mm. Order **Collembola**

ventral tube

20 Antennae and eyes absent; forelegs held upwards when walking . . . .   Order **Protura**
(proturans)

Long slender antennae **21**

Proturan, 1.5 mm. Order **Protura**

21 Three bristles at the end of the abdomen . . . . .   Order **Thysanura**
(bristle-tails)

Two long processes at the end of the abdomen. No eyes . . . .   Order **Diplura**
(diplurans)

Bristle-tail, 20 mm. Order **Thysanura**

Dipluran, 4 mm. Order **Diplura**

3 Key to the soil-inhabiting larvae of some families of beetles and two-winged flies

1 Body bearing no legs; head partially sclerotized, with simple eyes (Order Diptera) **8**
Body bearing legs; head completely sclerotized (Order Coleoptera) **2**

2 Body long and straight **3**
Body soft **6**

3 Legs long and well-developed **4**
Legs shorter or very short **5**

4 Cerci many-jointed; tarsi with one or two claws; carnivorous, active; in soil litter, under bark   Family **Carabidae** (ground beetles)

Two or three pairs of hooks on tergum of fifth abdominal segment; predacious; in wet meadows or river banks ....   Family **Cicindelidae** (tiger beetles)

cercus

Larva of ground beetle. Family **Carabidae**   Larva of tiger beetle, *Cicindela campestris*, 18 mm (Figure 6.1(a)). Family **Cicindelidae**

5 Body elongated, stiff, orange-brown; feed on grass roots ....   Family **Elatridae** (click beetles)

Body campodeiform; not stiff; tarsi without definite claw; cerci two-jointed; carnivorous, active; some parasitic ....   Family **Staphylinidae** (rove beetles)

cercus

Larva of click beetle, *Agriotes* sp., 15 mm. Family **Elatridae**   Larva of rove beetle (Figure 6.1(b)). Family **Staphylinidae**

6 Body curved and like a very large grub; inactive   **7**
Body not curved **8**

7 Most abdominal segments with two dorsal folds; in dung; some feed on roots ....   Family **Scarabaeidae** (dung beetles)

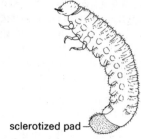

sclerotized pad

Larva of dung beetle. Family **Scarabaeidae**

Larva of stag beetle, *Lucanus* sp., 40 mm approx. Family **Lucanidae**

As above but with two sclerotized pads on either side of anus; in decaying wood . . . .   Family **Lucanidae** (stag beetles)

8 Thorax and abdomen of the same width **9**
Thorax much wider than head; body often woodlouse-shaped; in damp herbage or under bark . . . .   Family **Silphidae** (carrion beetles)

Larva of carrion beetle. Family **Silphidae**

9 Anal pair of spiracles obvious and surrounded by lobes; feed on plant roots, fungi, decaying wood   Family **Tipulidae** (craneflies)

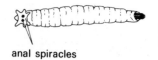

Larva of cranefly, *Tipula* sp. (Figure 6.2(a)). Family **Tipulidae**

anal spiracles

10 Body bearing rows of tubercles, often with bristles on each segment; in dung or grass roots . . . .   Family **Bibionidae** (march flies)

Body bearing no tubercles **11**

tubercles

Larva of march fly, *Bibio* sp., 15 mm (Figure 6.2(c)). Family **Bibionidae**

11 Body pear-shaped with pointed head; many are root pests of crops; some are saprophagous .   Family **Muscidae** (horse and stables flies)

Anal spiracles borne at the end of a pair of long, stalked processes; in dung or decaying organic matter ....   Family **Scatopsidae** (dung midges)

Body without conspicuous bristles; feed on fungal mycelia ....   Family **Mycetophilidae** (fungus gnats)

spiracle

Larva of muscid fly. Family **Muscidae**   Larva of dung midge. Family **Scatopsidae**

4 Key to the Order Acari

1 Terminal segment of pedipalp with a forked seta ventrically on its inner base (Figure a); stigmata usually with elongated peritremes, one on each side of the body, situated ventro- or dorso-laterally in the region of the coxal joints of any of the four pairs of legs (Figure b) ....   **Mesostigmata** (or Parasitiformes)

forked seta

pedipalp

peritreme

stigma

(a)

(b)

Mesostigmatid mite (a) pedipalp and (b) ventral view (only bases of legs shown). **Mesostigmata**

Terminal segment of pedipalp without forked seta. Stigmata usually without peritremes ...   2

2 Hypostome modified into a harpoon-like structure (Figure c); stigmata, one on each side of the body, situated anterior or posterior to coxae of fourth pair of legs; few soil representatives; ectoparasitic on vertebrates ....   **Metastigmata** (or Ixiodes)

hypostome

Gnathosoma of metastigmatid mite. **Meta-stigmata**

(c)

Hypostome not modified into a harpoon-like structure; stigmata situated on various parts of the body or absent; with or without pseudo-stigmatic organs **3**

3 Chelicerae chelate, rarely modified; pedipalps simple; tibia of first and second pair of legs each with long, whip-like seta (Figure d); body weakly or strongly sclerotized . . . . **Cryptostigmata**
(or Sarcoptiformes)

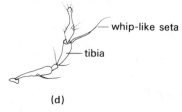

(d)

First leg of cryptostigmatid mite. **Crypto-stigmata**

Chelicerae and pedipalps usually strongly modified (Figures (e) and (f)); tibia of first and second pair of legs without whip-like seta; body usually poorly sclerotized . . . . **Prostigmata**
(or Trombidiiformes)

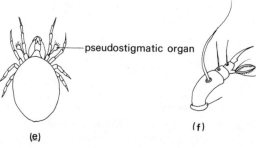

(e)

Prostigmatid mite, dorsal view

(f)

Pedipalp

Soil Micro-organisms

The micro-organisms in soil belong to many taxonomic groups of both animal and plant kingdoms. The micro-flora include fungi, bacteria, and actinomycetes as well as algae, while the protozoa and nematodes are the most important groups representing the micro-fauna.

In the following brief description of each of these groups the characteristics described are those of interest to the soil microbiologist and not necessarily typical of the group as a whole.

Viruses and bacteriophages

It is somewhat debatable whether viruses should be included as living organisms, since they can only exist and multiply inside living cells and are themselves non-cellular. Chemically, viruses consist of proteins and nucleic acid. Plant viruses living within plant cells, contain only RNA (ribonucleic acid) while those inhabiting bacteria usually contain only DNA (deoxyribonucleic acid). All viruses are obligate parasites, most being unable to survive for long outside the host. Some, like tobacco mosaic virus, can remain quiescent in the soil for several months. Many can be transmitted from one plant to another by larger soil organisms. The viruses causing raspberry leaf curl and peach yellow bud are probably transmitted by the nematode *Xiphinema* sp. for instance. By eliminating the vectors, the incidence of the disease can be reduced.

The group of viruses which parasitize bacteria are called bacteriophages (or simply phages) and when multiplying within the host cell, cause it to rupture. Some phages are specific to a single host species, others are more catholic in their host specificity.

Bacteria

Without doubt, bacteria are the most numerous organisms in soil and play an important role in many soil processes. There can be as many as 10^9 bacteria in a single gram of soil (or a live weight of over 890 kg/hectare). Under favourable conditions, bacteria multiply rapidly but this can only be maintained for short periods since the nutrients essential for growth and division are soon exhausted. Nevertheless, such rapid reproduction means that bacteria can compete successfully with other organisms by taking advantage of any new source of food.

The type of bacteria most commonly found in soil are rod-shaped and approximately 1μm wide and 2 to 3 μm long. Many possess flagella by means of which they can swim actively about in the soil. Often a sheath or capsule of complex chemical composition is secreted outside the actual cell wall (Figure 4.1). The capsule may protect the bacterium from being ingested by certain

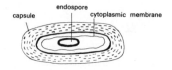

Figure 4.1 A typical capsulated bacillus

protozoa but the presence of large numbers of capsuled bacteria are thought to improve the crumb structure of a soil by binding particles of humus and mineral matter.

The ordinary bacterial cell is not resistant to high temperatures nor to desiccation, but some species are capable of forming resistant endospores which can withstand both heat and prolonged periods of drought.

Most soil bacteria require oxygen and their oxygen demands can soon create anaerobic conditions especially in waterlogged soils.

So far as their nutritional requirements are concerned, soil bacteria can be divided into two groups, the heterotrophs and the autotrophs. The latter obtain their supply of carbon from the carbon dioxide circulating in the soil and their energy either by the oxidation of inorganic substances or directly from sunlight. Heterotrophic forms use different organic substances including sugars, cellulose, organic acids, and hydrocarbons to supply their carbon and energy requirements. Their ability to decompose herbicides and pesticides prevents the build-up of these substances in the soil. By circulating important nutrients such as carbon, phosphorus, and nitrogen in the soil, bacteria play an essential role and more will be said about this in Chapter 9.

Actinomycetes

Actinomycetes superficially resemble fungi since in their vegetative stages they consist of fine branching filaments, about 1μm in diameter, although the similarity of simpler forms such as *Mycobacterium* to certain bacteria has led some authorities to consider them as highly evolved and complex bacteria.

Reproduction is by the fragmentation of the filaments sometimes forming dense colonies, into bacterium-like pieces, but in more complex species the mycelium is stable. Most species produce spores either singly or in chains (Figure 4.2).

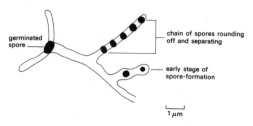

Figure 4.2 An actinomycete, *Streptomyces* sp.

Many actinomycetes can grow at high temperatures and saprophytic species play an important part in the fermentation of composts and manures. They have a distinctive odour when cultured, which may account for the earthy smell of soil. Others are important for their production of antibiotics. *Streptomyces griseus* produces streptomycin and *S. aureofaciens*, aureomycin. Yet others can be the cause of disease in humans, animals, and plants. To mention but a few of the pathogenic forms, *Mycobacterium tuberculosis* causes mammalian (both human and bovine) tuberculosis and *M. leprae*, leprosy in man. Common scab disease of potatoes is caused by *Streptomyces scabies*.

Fungi

Although of equal importance to bacteria in the contribution they make to soil processes generally, fungi are more plentiful in acid soils to which bacteria are less tolerant.

Fungi not only occupy a variety of habitats within the soil or connected with it but also differ greatly in size and mode of life. Some are active plant and animal parasites causing disease, many are saprophytic, feeding on a wide range of substances found in dead or decaying organic material, and others form close symbiotic relationships with plant roots.

Belonging to the class Phycomycetes are many common soil saprophytes such as species of *Mucor*, and also parasitic moulds. A description of one, *Pythium debaryanum* (Figure 4.3), which is responsible for the damping off of seedlings, must suffice. In this species the mycelium consists of long, irregularly branched, aseptate hyphae and under the microscope, active streaming of the cytoplasm along a hypha can easily be seen, food substances being carried in this way towards the growing tip or towards the developing reproductive bodies.

Pythium penetrates the tissue of the host seedling by enzyme digestion of the cellulose cell walls of the root, eventually growing through the host tissue and destroying it. As in most of the Phycomycetes, reproduction is both sexual and asexual. In the latter case aerial hyphae form a number of terminal sporangia on side branches. When ripe the contents of the sporangium pass into a vesicle breaking up into a number of zoospores. The wall of the vesicle ruptures to liberate the biciliate zoospores dependent on moisture for their germination and dispersal. In the case of cultivated seedlings attacks by *Pythium* are usually the result of over-watering. In the absence of a suitable host, *Pythium* can survive saprophytically, and under conditions of drought, sexual reproduction takes place within the tissues of the host or on aerial hyphae, involving the production of oogonia and antheridia which unite to form a resistant oospore. Thus the asexual spores are the agents of multiplication and dispersal while the sexual spores, with their store of foodstuffs, provide the means of survival.

Other Phycomycetes abundant in soil are obligate symbionts of higher plants, their germinating spores having only a limited growth unless the germ tubes come into contact with a plant root. Once this happens they penetrate the root cortex and develop an extensive mycelium within the corticle cells. This association appears to be truly symbiotic, the fungus depending for nourishment on the cell sap while the plant benefits as a result of the increased supply of mineral salts in solution conducted from the soil by the fungal

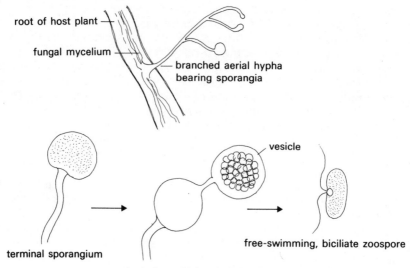

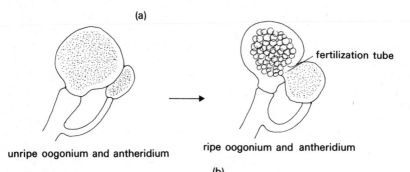

Figure 4.3 *Pythium debaryanum* (a) asexual reproduction and (b) sexual reproduction

mycelium. This type of association between fungus and root is a form of endophytic mycorrhiza occurring in a wide range of plants. Fungi belonging to the Basidiomycetes, form ectotrophic mycorrhiza in association with the roots of certain plants. These associations will be referred to again in Chapter 9.

Other Basidiomycetes are common in soil forming extensive mycelia, their large fruiting bodies being the conspicuous, above-ground toadstools of various kinds. The destructive honey agaric, *Armallaria mellea*, destroys the roots of many plants including trees. The black aggregated strands of mycelium, called rhizomorphs, can spread for several metres through the soil from their attachment to a decaying tree stump or root from which they draw food. In this way the fungus can spread to attack plants some distance from the original victim and is consequently difficult to eradicate.

The third and largest group of fungi, the Ascomycetes, produce vast numbers of asexual spores, while they appear to have no sexual stage when cultured.

Because of their method of producing asexual (imperfect) spores, these groups are often termed Fungi Imperfecti. The numerous species of *Penicillium* are examples, many of which produce antibiotics which can be isolated from cultures. The spores germinate to form a network of septate hyphae, penetrating the surface soil around soil particles and leaf mould from which they obtain nourishment. The mycelium produces conidiophores from which the chains of spores arise (Figure 4.4).

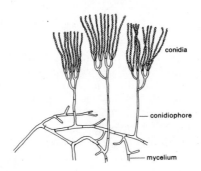

conidia

conidiophore

mycelium

Figure 4.4 *Penicillium* sp. Mycelium with conidiophores producing chains of conidia (spores) highly magnified

Other members of the Ascomycetes are the yeasts of which there are many species. They are present in all soils although a particular soil generally possesses a typical yeast flora not found in other places. The well-known yeasts associated with vineculture are examples, each species being responsible for the fermentation processes connected with the production of a particular wine. Yeasts are single-celled fungi and those responsible for the fermentation of grapes occur on the vine leaves and become washed by rain into the upper layers of the soil in which they multiply and get blown once more on to the leaf surfaces.

Soil fungi, so briefly described here, can make use as food of a wide range of organic matter. They are, for instance, the organisms responsible for the decomposition of lignin in wood as well as cellulose, making these and other substances, available to a variety of soil organisms. Not only are some fungi capable of parasitizing plants but others attack and parasitize soil animals such as nematodes and play an important part in controlling nematode species which parasitize plants. The ability of many fungi to produce resistant spores enables them to survive drought and frost.

Algae

Left undisturbed, the surface of compacted soil will often have a green, slimy appearance. This is due to an abundant growth of algae. They differ from the other micro-organisms already mentioned, in the possession of chlorophyll and therefore have the ability to synthesize sugars from carbon dioxide and water using sunlight as the energy source. Because they require light for their autotrophic nutrition, algae are found in greater abundance at or just below the surface. Heterotrophic forms, dependent upon organic compounds, do occur at depths of up to 20 cm below the surface, but are never as numerous as the autotrophs.

Because they can synthesize their own food, algae are often the first colonizers of bare ground in places such as volcanic soils or burnt land remote from the possibility of seeding by higher plants. Economically, blue-green algae can play an important part in the flooded surface soils of paddy fields by supplying the rice plants with nitrogen by direct nitrogen-fixation.

Protozoa

Three classes of protozoa, which are single-celled animals, have representatives in soil – the rhizopods, flagellates, and ciliates (Figure 4.5). The latter two are really free-swimming. The flagellates which are 3 to 10μm in length, possess one or more flagellae while the ciliates are covered with many cilia often arranged in bands. Both flagellates and ciliates are confined to the film of water surrounding the particles of soil. They feed on bacteria as well as upon other protozoa, but many contain green pigments and can feed like algae. Thus species such as *Euglena viridis* are often classed as algae.

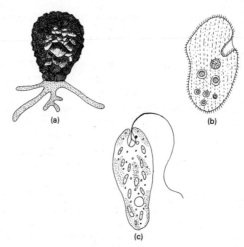

Figure 4.5 Some soil protozoans (a) *Difflugia*, 100 μm, a rhizopod, (b) *Colpidium*, 55 μm, a ciliate, and (c) *Euglena*, 50 μm, a flagellate

The rhizopods or amoebae are larger, predatory species being as much as 3 mm across. Some are naked blobs of protoplasm which can change their shape, putting out protoplasmic 'arms' or pseudopodia to engulf their prey. Others, sometimes called thecamoebae, like *Arcella* and *Difflugia*, have rigid shells composed of chitin or silica from which fine protoplasmic threads emerge. Most amoebae prey upon bacteria and although they are a definite element in the control of numbers, they do not impair the function of bacteria in the soil ecosystem. It is also true to say that not all bacteria are acceptable as food and some amoebae are saprozoic, feeding on decaying animal and vegetable material in the soil.

Most soil protozoa are capable of encystment in unfavourable conditions. The cysts can tolerate a wider temperature and humidity range than the active forms and are readily dispersed by wind.

Nematodes

Among all the groups of soil fauna the nematodes are probably the most ubiquitous being found wherever there is moisture. They are associated with all types of fresh water and in soil are present in the moisture film around soil particles. The best known are those which are parasitic on animals and plants but probably the majority of species are free-living.

Most soils contain vast numbers of nematodes or eelworms, populations of 10–20 million/m² being not uncommon. They are most abundant in the top 5 cm of soil, especially near roots or other plant material, although their diets vary from diatoms, bacteria, and algae to the juices of plants or even other nematodes.

Most free-living species are less than 1.0 mm long and have elongated, cylindrical bodies encased in a tough cuticle. Figure 4.6 shows the structure of a typical eelworm common in most soils. Identification, which is difficult for anyone who is not a specialist, is based on the structure of the buccal capsule.

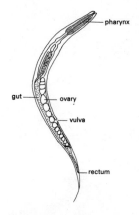

Figure 4.6 *Rhabditis* sp., O 1.2 mm, a free-living nematode common in soil

Free-living members of the genus *Rhabditis* which are saprophytic and bacterial feeders, have narrow mouths and a long muscular pharynx. Stylets within the buccal cavity, are present in all species parasitic on plants, the stylets being used for piercing plant tissue. Predacious species are characterized by the development of 'teeth' surrounding the buccal cavity (Figure 4.7).

Their various feeding methods make nematodes well adapted to the transmission of viruses from one plant to another and as efficient vectors of such diseases are rivalled only by aphids.

Some species such as *Cheilobus quadrilabiatus* produce larvae which attach themselves to staphylinid beetles by means of which they are transported. Such phoretic associations with insects are quite common.

Of great economic importance as crop pests are the eelworms which attack

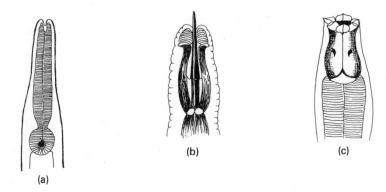

Figure 4.7 Head and buccal capsule of (a) a saprophytic nematode, (b) a nematode parasitic on plants showing stylets used for piercing plant tissue, and (c) a predacious species with 'teeth' surrounding the buccal cavity

Figure 4.8 Potato cyst eelworm, *Heterodera rostochiensis*. Cysts on potato roots. (*Photo by courtesy of C. C. Doncaster.*)

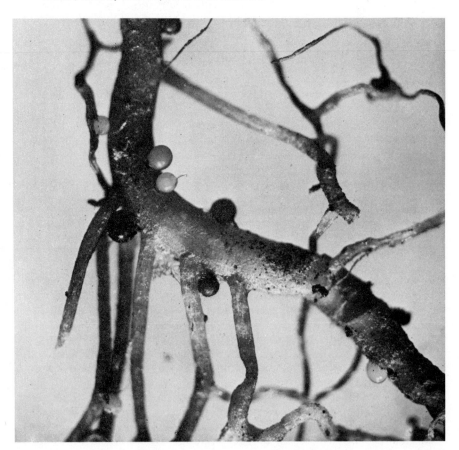

plant roots, stems, foliage, and so forth. In most cases at least, part of the life cycle takes place in the soil. After multiplying within the host plant, they may migrate quite long distances through the soil often assisted by run-off from banks or hillsides. Migration will take place also on the death of the host plant but under favourable conditions, they will adopt a resting stage. The Heteroderidae are cyst-forming eelworms, *Heterodera rostochiensis*, the potato root eelworm, is one of the best known. The cysts (Figure 4.8), about 1 mm in diameter are, in reality the swollen bodies of dead female worms containing a large number of eggs which will only hatch under the influence of exudates from the roots of the host plant. The minute 'larvae' travel through the soil swimming in the water film of the soil pores, until they reach the roots of a fresh host. Here they pierce the root cortex with their mouth stylets and thereafter mature within the host plant.

Although some species provide an important source of food for other members of the soil community, notably mites, the general opinion seems to be that nematodes do not contribute significantly to the break down of organic material in the soil, but as bacterial feeders they may well have an indirect effect on decomposition.

The distribution of micro-organisms in the soil

In soil with a moderately uniform texture, the highest concentration of micro-organisms occurs in the first few centimetres below the surface, numbers decreasing rapidly with depth. In soils where there is a well-developed series of horizons or in a podzol profile with a well-marked B-horizon enriched with humus, the distribution is quite different and there is often a marked increase in numbers in the lower layers.

In studying the vertical distribution of micro-organisms, it becomes evident that many species and especially fungal spores, can be washed downward by rain and this is particularly likely in open, sandy soils.

We have already mentioned in Chapter 1, the possibility of an increase in anaerobic conditions in certain soils and under certain conditions. Such an increase will effect the vertical distribution of micro-organisms. The number of bacteria capable of multiplying under these conditions increases with depth down to approximately 30 cm, but below this depth numbers begin to decrease. The growth rate of fungi, on the other hand, is less affected by anaerobic conditions.

Moisture, too, has its affect upon the vertical distribution of organisms. The litter layer, especially in woodland, can be quite dry to the touch and those micro-organisms which inhabit the litter must survive such dry conditions either by migrating to the moister layers beneath or by the production of resistant bodies of one kind or another.

In this brief description of the distribution of micro-organisms, no account has been taken of the special conditions created by the presence of plant roots. The importance of the *rhizosphere*, or soil which is modified by root activity, is discussed in Chapter 9.

5 Soil Arachnids and Other Small Animals

We now turn to some of the small animals in the soil, most of which are larger than 200 μm. Nevertheless, the majority are small enough to require a lens in order to see their structure properly.

Arachnida

In general the arachnids can be described as predatory arthropods occurring, for the most part, among aerial vegetation, on the surface of the soil or in the litter. Spiders (Araneida), harvestmen (Opiliones), false scorpions (Chelonethi), and mites (Acari) all belong to the class Arachnida and, in addition, there are two groups, the true scorpions (Scorpiones) and the sunspiders (Solifugae) which are tropical or semi-tropical.

Some groups of spiders, notably the trap-door spider, wolf spiders, and purse spiders, are closely associated with the soil community and undoubtedly prey on insects and other small arthropods. One colourful and easily recognized spider is the large, fierce-looking *Dysdera crocata* which has a rusty-red cephalothorax and bright yellow abdomen. By day these spiders are usually to be found in silken cells beneath stones, emerging to hunt at night. Bristowe (1958) describes how this spider feeds almost exclusively on woodlice. When it encounters its prey, it turns its cephalothorax on one side thus bringing its huge and powerful chelicerae into a position for seizing the woodlouse between its fangs (Figure 5.1). Bristowe showed that there was an order of preference for the different genera of woodlice, *Porcellio*, *Oniscus*, and *Armadillidium* being more distasteful than *Philoscia* and *Platyarthrus*. This was connected with the relative development of lobed glands secreting a distasteful fluid when the woodlice are attacked.

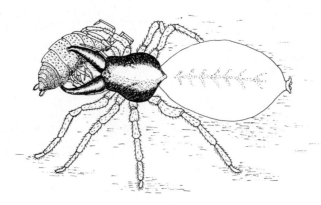

Figure 5.1 *Dysdera crocata*, 14 mm, seizing a woodlouse

There are many other species of spiders living on the surface or in soil crevices which are voracious predators of many small soil animals. The extent to which these spiders exert an influence on the soil community must be considerable, although few figures are available of the magnitude of spider predation.

Harvestmen without doubt also find much of their prey among the soil fauna such as carabid larvae and small centipedes. Above the litter, insects form their principal diet. Some species are parasitized by larval mites which are bright red and can be found attached to the body and legs of the harvestmen. Large infestations can cause the death of the host. Harvestmen are nocturnal in their habits, enjoying damp surroundings, and are mostly to be found in the litter layer.

Chelonethi

False scorpions, which closely resemble true scorpions in everything but size, are also found in moist vegetation on the surface of the soil, particularly amongst forest litter, although they are never very abundant in temperate soils. They are best found by examining samples of dry oak litter under a low-power lens, or they can be extracted from litter using one of the dry funnel methods (see page 80).

False scorpions are unmistakable due to their palps which resemble lobster claws (Figure 5.2). In Britain the largest species is only about 4 mm. Those

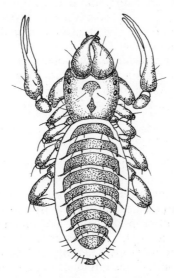

Figure 5.2 False scorpion, 2mm. From beech litter

which inhabit soil or bark crevices have no eyes but in other species there are two or four eyes on the anterior part of the cephalothorax. The number of eyes and the number and position of the tactile hairs, or *trichobothia*, are important diagnostic characters. A key for identification of the common British species can be found in Cloudsley-Thompson and Sankey (1961).

False scorpions are exclusively carnivorous, feeding on living or recently killed prey such as small flies, arachnids, collembolans, and symphylids. It is doubtful if food is actively hunted. It is more likely that the false scorpion lies in wait, becoming aware of its prey only when it brushes against the very sensitive tactile hairs. The role of the poison glands present in the palps is not clear, for observation of captive animals shows that sometimes the prey is paralysed immediately but is often conveyed to the chelicerae whilst still alive and struggling. Both before and after a meal a false scorpion can sometimes be seen cleaning the chelicerae and mouthparts. This is important because digestion of the prey is external by means of enzymes. It is therefore essential that the grooved mouthparts should be kept free of solid particles so that the pre-digested body juices of the prey can be sucked up.

False scorpions can probably hold their own against species of the same size as themselves. Their secretive habits and their poison-bearing pedipals mean that they have few enemies although there are records of them being eaten by ants. Their own position in the detritivore-predator food web is probably as tertiary consumers and because of their low numbers and scattered distribution, they can have little effect upon detritus-feeding soil animals.

Male false scorpions, like true scorpions, perform a kind of ritual mating display. Females of some species carry their eggs around with them, others construct nests in which they seal themselves up. When the nymphs hatch from the eggs they resemble their parents in all but size and may remain attached to the female's body for some days. They moult passing through three nymphal stages – protonymph, deutonymph, and tritonymph.

The whole fascinating saga of the courtship, life history, and feeding habits of false scorpions can be observed by keeping them in small chambers in which the atmosphere is kept constantly moist. Different food can be offered in the shape of small flies from the litter layer, collembolans, and proturans. In fact their biology presents many interesting features, not least of which is their phoretic behaviour which enables them to be distributed from one area to another by attaching themselves to small flies and hymenopterans.

Acari

Mites, which are especially abundant in organic woodland litter, are the commonest representatives of the meso-fauna in soil. They are by far the most important of the soil-inhabiting arachnids, often being present in astronomical numbers. Those which hit the headlines are mostly species parasitic on man or his domestic animals and crops and they are therefore of great economic importance. One example is the red spider mite, *Tetranychus urticae*, an important pest of greenhouse crops. This mite and its controlling predator will be described when we consider the influence of man on the soil community in Chapter 11. Within the context of this book however, we are concerned mostly with the free-living mites in the soil although a few are parasitic.

In mineral soils very few mites are found far beneath the surface except the very small species. Besides occupying the litter and surface layers of soil, mites are to be found in such diverse habitats as tree holes, lichen-encrusted rocks, moss, dung, and tidal debris.

The soil acari (Figure 5.3) are represented by members of four orders: the

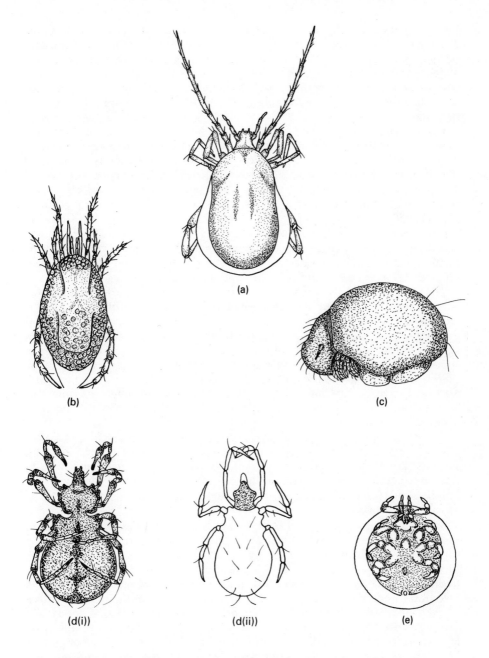

Figure 5.3 Some mites to be found in organic woodland litter: (a) *Pergamasus* sp., 2 mm. A mesostigmatid mite from larch litter. (b) Cryptostigmatid mite, 2 mm. From beech litter. (c) Oribatid armadillo mite, 1.5 mm. From larch litter. (d) Oribatid mite; (i) adult, 1.5 mm, ventral view and (ii) hexapod larva, 0.5 mm. From beech litter. (e) Mesostigmatid mite, 1 mm, ventral view. From beech litter

Cryptostigmata, Prostigmata, Mesostigmata, and Metastigmata (see Chapter 3, Key 4). Most of the Cryptostigmata are vegetarian, feeding on dung and decomposed vegetable matter. They are slow-moving and have blunt mandibles which are adapted to biting and crushing. A few species are carrion-feeders and coprophagous. The other three orders include species which may be predacious, phytophagous, mycetophagous, necrophagous, coprophagous, or parasitic.

The predatory species feed on a variety of animals including collembolans, proturans, nematodes, and enchytraeids. Some which live in dung or compost heaps, eat the eggs of the housefly. Food may well include what is available so that in woodland and forest soils, cryptostigmatic mites may be the diet of carnivorous species. When other food is scarce, some mites such as *Pergamasus* sp., may turn cannibalistic. They probably do not ingest solid food but perform external salivary digestion.

Mites possess a variety of sensory organs by means of which they can appreciate minute changes in their environment. For instance, movement of tactile setae stimulate sensory cells at the bases of the setae. Some species possess chemoreceptors and olfactory and humidity receptors.

The Oribatei, or 'beetle mites', found in soil litter and in moss cushions, are characteristic. Their bodies are typically partially sclerotized and the 'armadillo' oribatids can actually fold the fore part of the body against the hind part, making themselves almost spherical. In this way they protect themselves against desiccation or flooding of the soil as well as against predators such as staphylinid beetles and predatory mites. All the Oribatei are very important for their contribution to the breakdown of organic matter in the soil and we shall refer to this again in Chapter 9. Many species feed on fungal mycelia and spores, some on damp wood, and others burrow into fallen twigs and pine needles.

The general pattern of the life cycle of the soil mites, of which the Cryptostigmata form typical examples, starts with the emergence from the egg of a hexapod larva which then undergoes three nymphal changes (protonymph, deutonyph, and tritonymph) before becoming adult. However, there are variations in this life pattern and in some groups we have only scant information about the life histories.

Spermatozoa are introduced by the male directly into the receptaculum of the female by means of a sclerotized aedeagus or, in some groups, by means of a spermatophore which is transferred by the male to the female by the chelicerae. In the Prostigmata and Cryptostigmata the males attach stalked spermatophores to the substrate which are later taken up by the female. Under favourable conditions, the hexapod larvae may hatch from the eggs within a week, from where they are often laid in the cast nymphal exuviae. After a short feeding period the larvae become inactive and moult changing to an eight-legged protonymph. Feeding recommences and there is a further moult to become a deutonymph. The number of nymphal stages varies and in some species the protonymph is replaced by a resistant anabiotic 'hypopus' which is immobile and can be dispersed by wind or by becoming attached to other animals. In many groups the occurrence of numbers of larvae or nymphs can cause fluctuations in the population. This is particularly true of the Cryptostigmata which occur in greater numbers and therefore make a major

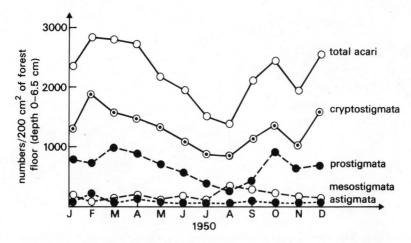

Figure 5.4 Seasonal fluctuations in the acarine fauna of litter and soil in a Sitka spruce plantation, based on densities. (*From Evans* et al. (1961).)

contribution to the fluctuation of the mite population of a soil throughout the year. Studies made by Evans *et al.* (1961) (see Figure 5.4) showed that there was a marked decline in the total number of mites in summer when there was a preponderance of adults, rising to a peak density in spring when the numbers of immature forms were at a maximum. It would seem that the long-lived protonymph is in most species the over-wintering stage with deuto- and trito-nymphs developing in spring. The high mortality of these nymphal stages is probably a factor contributing to the decline in numbers during summer.

Collembola

The Collembola or spring tails, are primitive, wingless insects belonging to the Apterygota. They are called 'spring tails' because those species which live near or on the surface of the soil possess a springing organ by means of which, for their size, they can make considerable jumps. The springing organ is bifid and when not in use is folded into a groove on the underside of the abdomen.

The Collembola can be divided into two groups, those which live on or near the surface and those which live beneath it (Figure 5.5). This is not strictly a taxonomic division since many species normally associated with the surface can be found beneath it and vice versa. However, broadly speaking, the surface species possess springing organs, compound eyes, and elongated antennae. Many breathe by means of tracheae. None are more than a few millimetres in length and those normally inhabiting the deeper layers are even smaller, a characteristic connected with the size of the soil pore spaces in which they live and move about. Subterranean species have no tracheae or springing organs, a permeable body covering, short antennae, and simple ocelli. All collembolans possess a ventral tube on the first abdominal segment, which can be everted or retracted and can also be used to attach the animal to the substratum.

Like the mites, the Collembola are extremely numerous. They show varying degrees of tolerance to different environmental factors such as the soil

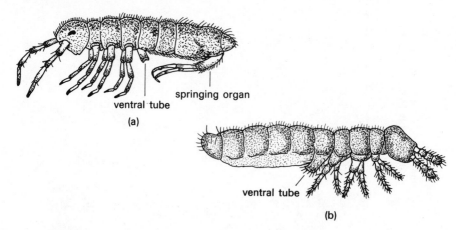

Figure 5.5 Two species of soil Collembola: (a) *Entombyra* sp., 2 mm. From larch litter. An epidaphic, tracheate species with eyes and springing organ. (b) *Onychiurus* sp., 1.5 mm. From the deeper layers of larch litter. A non-tracheate, blind species lacking a springing organ

structure and its type, the presence of micro-flora and moisture content, not to mention the soil pore size which, as we have seen, can be a limiting factor. Indeed the presence or absence of a species could be an indication of micro-habitat conditions. So far as soil micro-flora are concerned, it is fairly certain that some species live in the gut of collembolans as symbionts, probably assisting in the digestion of plant material. More work needs to be done on this relationship, however, before it can be regarded as a factor affecting the distribution of the species of collembolans which undoubtedly use soil micro-flora as food. The distribution of fungi upon which these species feed, must also be a controlling factor in their own distribution, especially as the fungal and collembolan populations will be competing for the available moisture.

In conclusion it is probably true to say that the composition of the species of Collembola in any habitat is determined by a number of factors and not by any single factor.

Since collembolans are unable to burrow, they must perforce use the soil pores in which to move. But these spaces will be smaller in regions where decomposition is taking place and where food is more abundant. Also, because of the greater amount of vegetable matter present, the soil is less prone to dry out. Vertical distribution is therefore governed by several factors – the size of the species and pore space, moisture content, and available food. In grassland, where there is little surface litter or a well-defined fermentation layer, the numbers of collembolans decrease progressively with depth.

Before leaving the question of distribution patterns, reference should be made once more to the non-tracheate species. In these spring tails, the ventral tube and the general body surface is used for gaseous exchange. The discontinuous wax layer of the cuticle and the hydrofuge hairs confer some degree of protection against desiccation as well as against flooding, for the hairs trap a film of air and this permits cuticular respiration to continue even under conditions of extreme moisture. The evolution of a tracheal system in the more advanced species allows for a greater waterproofing system and therefore permits a degree of independence making possible the colonization of more exposed habitats without the danger of desiccation or suffocation.

The ability to perform vertical movements in the soil is of advantage to the smaller species which, although usually associated with lower levels, may nevertheless range through the profile. These vertical migrations mean that the animals can move upwards to moister conditions in spring and autumn and avoid drought and extremes of temperature by moving downwards in summer and winter. In this way they can take advantage of optimal conditions of temperature and moisture.

The mouthparts of collembolans vary from one group to another. Broadly speaking those which chew their food possess well-developed mandibles and a molar plate capable of rasping hard plant material. Biting forms have strong, toothed mandibles but no molar plate while in sucking forms the mandibles are reduced to stylets for piercing and sucking juices. Some specialist carnivorous species prefer rotifers, proturans, and tardigrades while phytophagous forms may select one species of plant as their food. However, although it would make a tidy classification of types to correlate mouthparts with the type of food selected, in point of fact many species consume a wide variety of organic food according to what is available.

In most Collembola development from egg to adult occurs during one year, some producing two or more generations within that time. Sperm is usually transferred indirectly from male to female by the male who deposits either a spermatophore or a free droplet on the substratum which is later taken up by the female. Development and hatching of the eggs depends on the species and on the temperature of the soil. The first instar hatching from an egg is six-legged and resembles the adult in all but size and pigmentation. Thereafter, the young moult several times giving rise to successively larger more darkly coloured individuals until the final instar which is deemed to be that at which maximum size is attained. This is, however, obviously difficult to determine and there is also a variable number of instars according to species.

The production of large numbers of juveniles gives rise to population peaks at certain times of the year and these will be influenced by local environmental factors, particularly temperature.

The declines in population, alternating with the peaks, occur markedly in January and July in northern latitudes because of increased mortality largely due to predation by mesostigmatid mites.

Rotifera and Gastrotricha

Rotifers or wheel animalcules, are fresh water animals never exceeding 2 mm in length, but one order, the Bdelloidea, are well represented in the upper layers of the soil and especially in mossy litter. Inhabiting the water film in the soil they can progress by means of the wheel organ on the head or by a looping action in which the foot, with its mucus glands, participates. The Bdelloid rotifers (Figure 5.6) are capable of contracting their bodies in such a way that they become unrecognizable. Their integument exudes a fluid which stiffens into a gelatinous sheath surrounding their bodies on which granules and flakes of excrement collect. These sheaths give protection against desiccation.

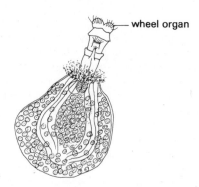

Figure 5.6 *Habrotrocha tridens excedens*, 1 mm approx. A Bdelloid rotifer in its nest of mucus and faeces, with collar of dirt

Parthenogenetic eggs are laid inside the sheath, males being unknown amongst the members of this group.

Like other rotifers, the Bdelloids are vortex feeders: cilia surrounding the wheel organ beat continuously and create a current which conveys organic particles to the mouth to be ingested. In drought conditions these rotifers assume an anabiotic existence becoming withdrawn, shrivelled, and able to withstand a wide range of temperature. In this condition they can be dispersed by wind currents and are thus capable of surviving longer than their usual lifespan of a few months.

Mention must be made of the little-known 'Hairy backs' which are now placed in a separate class, the Gastrotricha. Although insignificant in size, the largest being less than 0.5 mm, they are quite common and live in the soil water film like ciliate protozoans, which they closely resemble. They are multicellular and creep in the film by means of spiny cilia. *Chaetonotus* sp. (Figure 5.7(b)) is one of the gastrotriches most commonly encountered.

Tardigrada

Tardigrades are a group of very small animals with a relatively complex body structure, sometimes called 'Water bears' because, when highly magnified, they have a superficial resemblance to a bear (Figure 5.7(a)). In common with rotifers and some ciliates, water bears creep in the soil water film or in water films surrounding moss cells, for they are essentially animals inhabiting damp places. They are considered to be arthropods because of their four pairs of legs and chitinized body covering. However, since they do not closely resemble any particular group of arthropods, they have been placed in a class by themselves.

Some species of tardigrade are found in the soil itself feeding on leafy remains, but the majority live in moss cushions where they feed on the cell sap by piercing the cells with evertable mouth stylets and sucking out the sap by means of a muscular pharynx. The females lay thick-walled eggs inside their cast skins. These anabiotic eggs are resistant to desiccation and may be blown about until the return of moist conditions.

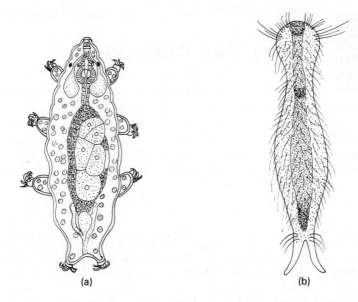

Figure 5.7 (a) *Macrobiotus* sp., 400 µm, a tardigrade (water bear) and (b) *Chaetonotus* sp., 300 µm, a gastrotrich (hairy back)

6 *Larger Soil Animals*

Arthropods

The arachnids and primitive wingless insects (Apterygota) were described in the last chapter. We now come to the other arthropods found in the soil: the insects, crustaceans, and myriapods.

Insects

Many insect orders have their soil representatives. Their distribution is often uneven and, by comparison with other soil organisms, their numbers are small.

For the purposes of this book we shall be concerned only with those which make a significant contribution to the soil community, namely the Coleoptara (beetles), Diptera (two-winged flies), Hymenoptera (ants, bees, and wasps), Lepidoptera (moths and butterflies), and Orthoptera (crickets and grass-hoppers). The Isoptera (termites) are omitted since, apart from two genera, they are not found in temperate soils.

Coleoptera

Beetles, both as adults and larvae, are probably the most diverse in habits and structure of all soil-inhabiting insects. The large number of beetle species is witness to their success as a group, one reason for which is their ability to adapt to a wide variety of habitats and food. So many species form an active part of the soil community that there is space for but a few examples which illustrate some of their adaptations and contributions to life in the soil and litter.

The ground beetles (Carabidae) with their long legs are well adapted to running over the surface of the soil. They prey on a variety of small animals including mites and other insect larvae. The larva of the handsome metallic green tiger beetle, *Cicindela campestris*, digs a pit in which it lies vertically with its head at ground level waiting for insect prey to come within reach of its powerful jaws (Figure 6.1(a)). This method of closing the entrance to its burrow with its head is termed *phragmosis*, a phenomenon shared with certain species of soldier ant.

The staphylinids or rove beetles (Figure 6.1(b) and (c)) are typical soil coleopterans. Some are carrion feeders and are quite large, like *Necrophorus vespillo*. *Lomenchusa* is a small species whose larvae live in the nests of the ant, *Formica sanguinea*. The larvae have tufts of hair surrounding glands which produce a secretion much beloved by the ants, who feed on both the larvae and adult beetles despite the latter being capable of finding their own food. Kevan (1962) mentions that when danger threatens the ants rescue their 'guests' before their own young.

The dung beetles (Scarabidae) include the dor beetles of which *Geotrupes* (Figures 6.1(d)) is a common example. These beetles emerge at night from the

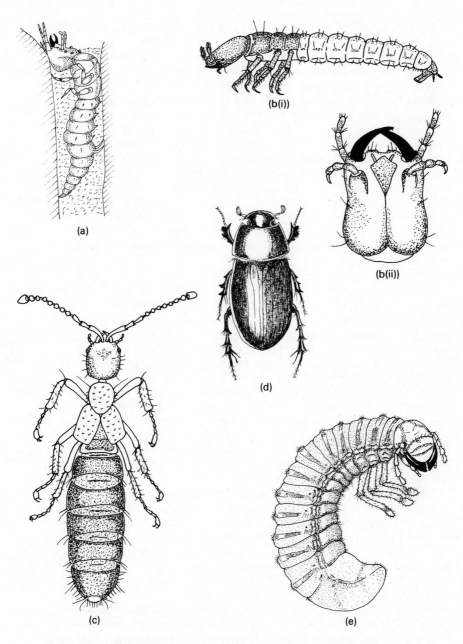

Figure 6.1 (a) Larva of tiger beetle, *Cicindela campestris*, 18 mm, in its burrow. Note the projecting jaws and the arrangement of the ocelli on top of the head, permitting vision in various directions. (b) (i) Larva of staphylinid beetle, 4 mm. From larch litter; (ii) Head showing powerful jaws. (c) Staphylinid beetle, 8 mm. From larch litter. (d) Dor beetle, *Geotrupes* sp., 12 mm. The broad head and the expansions on the legs assist burrowing. (e) Larva of cockchafer, *Melalontha melalontha*, 40 mm

soil and dig a deep shaft, some 30 cm or more below a cowpat. Working rapidly they drag down the manure from above, several beetles being capable of burying a complete cowpat overnight. They use the dung thus interred, as food during the day or in bad weather. In this way a great deal of organic matter can be incorporated in the soil beneath a meadow.

Other beetles or their larvae make an important contribution to the soil by their decomposition of organic material in various ways such as feeding on carrion or decaying wood which is returned to the soil in their excrements. Some of the larger beetle larvae, like those of the stag beetle, *Lucanus cervus*, and the cockchafer, *Melalontha melalontha* (Figure 6.1(e)), can consume vast amounts of wood and damage tree roots by their feeding activities. These larvae remain subterranean for several years since they derive only a small amount of nourishment from wood and growth is consequently slow. *Agriotes* sp., the click beetle, are also serious pests as larvae, taking four years to mature. They feed on grass roots, but ploughing reduces their natural food and if corn is then sown, they will turn their attentions to the roots of that crop.

Diptera

So many flies have soil-dwelling larvae that only those most frequently encountered can be mentioned here. The larvae are usually confined to damper soils such as the fermentation and litter layers of forest soils, compost heaps, and dung. Most are not able to burrow very much and therefore depend on existing soil crevices or making a passage through loose litter.

Tipulids are crane flies with larvae commonly called leather-jackets (Figure 6.2 (a)), abundant in arable soils. By feeding on plant roots, they cause damage to crops. Like the wood-feeding insects, they are slow to mature and may take two years before pupating. The pupae have spines which enable them to wriggle to the surface just before the adult is due to emerge. On a warm evening many thousands may be emerging at the same time, a phenomenon often marked by swallows and martins hawking low over a meadow. Larval fungus gnats (Mycetophilidae) are restricted in diet to fungal mycelia, the adults emerging in the litter (Figure 6.2(b)).

Many dipterous larvae are coprophagous, feeding on vertebrate dung. The larvae of march flies, *Bibio* sp. (Figure 6.2(c)), moth flies, *Psychoda* sp., dung midges, Scatopsidae, and filth flies, *Fannia* sp. by their dung-feeding activities help in the faunal succession and the eventual products of decomposition being incorporated in the soil. In fact the ecology of dung pats can be the basis of interesting projects, for a community of dipteran larvae and their predators often exists in a single dung pat. *Fannia* sp. (Figure 6.2(d)) inhabits the nests of bumblebees, feeding on their dung.

Blowfly larvae, *Calliphora* sp., feed on carrion by pouring extra-digestive juices on the food which is then sucked up. Appearing as some of the first decomposers of carrion, they are important as scavengers.

Hymenoptera

The numberless kinds of bees and wasps which are fossorial, excavating nests in soil, cannot really be counted as true members of the soil fauna, although one family of solitary bees, the Andrenidae, by constructing tortuous underground nests certainly break up and aerate the soil. Neither bees nor wasps, however,

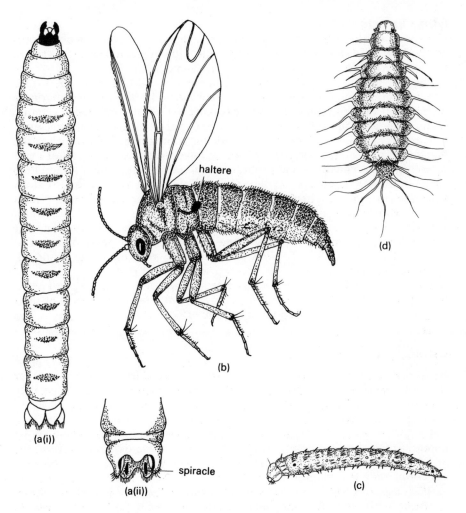

haltere

(b)

(a(i))

(a(ii))

spiracle

(c)

(d)

Figure 6.2 (a) (i) Tipulid larva, 6 mm, ventral view; (ii) Dorsal view of anal segment. (b) Fungus gnat (Mycetophilidae), 3 mm. From heath moorland. (c) Larva of march fly, *Bibio* sp., 15 mm. (d) Larva of filth fly, *Fannia fuscula*, 6 mm

are as important as ants which are often the pioneers in burrowing and breaking up new sites, particularly on banks and under stones where the galleries of the underground nests can be extensive. In making these nests, ants break down the soil to a fine powder and also bring up soil from lower levels. The wood ant, *Formica rufa*, collects large piles of plant debris for its woodland nests above ground. With the assistance of mites and collembolans, this debris is eventually converted into organic matter which is incorporated in the soil. Ants undoubtedly improve the crumb structure of soil and are also responsible for the translocation of soil particles beneath the surface.

Other insects

Cutworms are the larvae of noctuid moths and are pests feeding on plant roots and underground shoots, but most of the other lepidopterous larvae that live in the soil are not of much importance in the community.

Earwigs, belonging to the order Dermaptera, should be mentioned since although some are true soil denizens, they are nevertheless only periodic members of the soil fauna, frequently leaving the soil to hunt above ground. Some are predacious, others feed on decaying vegetation and fungal mycelia.

The true plant bugs, Hemiptera, include the root-feeding aphids such as the bulb and potato aphid, *Rhopalosiphoninus latysiphon*, a serious pest of these crops which has above-ground generations on surface vegetation. Around the Mediterranean and other warm regions, cicadas, whose larvae spend up to four years underground, are important crop pests which excavate extensive underground tunnels (see page 72).

Lastly, there are many orthopterous insects which spend a large part of their lives in the ground. These include the young stages of grasshoppers and crickets. The mole cricket, *Gryllotalpa gryllotalpa*, common on the continent, constructs underground burrows and is a serious crop pest.

The importance of these insects, apart from many of them being plant pests, is really the effect of their burrowing activities and the accumulation of their faeces upon which coprophagous species feed.

Crustacea

Isopod crustaceans are mostly marine, but woodlice (Figure 6.3) are familiar terrestrial animals found amongst rotting wood, in litter, and are especially common in compost heaps where they feed on a variety of dead and decaying organic matter. Woodlice lack an epicuticular wax layer and therefore lose water rapidly in dry conditions. For this reason, moisture is probably the most important factor in influencing their distribution. *Armadillidium vulgare*, for instance, can tolerate drier conditions than *Porcellio scaber*, while *Philoscia muscorum* requires still moister conditions (Figure 6.4). Rates of transpiration vary with species which may be correlated with differences in cuticular structure which probably becomes increasingly more permeable with a rise in temperature up to a point, varying with species, where the waterproofing barrier becomes destroyed altogether. None of the terrestrial isopods are completely independent of moisture because they breathe by means of modified gills requiring the presence of damp air. Their survival depends more upon behaviour and Cloudesly-Thompson (1952) showed that *Oniscus asellus* exhibits a photonegative response which increases with darkness. If conditions become drier, the animals become photopositive which enables them to move to damper regions where they once more become photonegative. Interesting though the realms of isopod behaviour and physiology are, they have only an indirect bearing on the subject of this book and the reader who wants to pursue the subject further is referred to Edney (1954) and for a general account of *O. asellus* and for a key to the genera of British woodlice to Darlington and Leadley Brown (1975).

In places where there is an accumulation of vegetable matter and where woodlice are abundant, they may be important in the breakdown of this

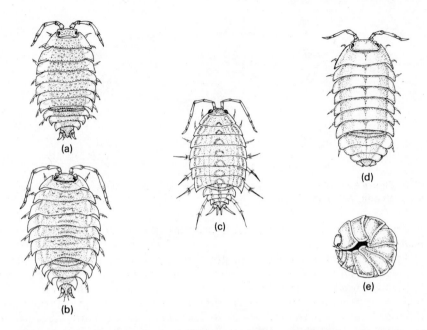

Figure 6.3 Woodlice commonly found in surface litter: (a) *Porcellio scaber*, 12 mm, (b) *Oniscus asellus*, 13 mm, (c) *Philoscia muscorum*, 8 mm, (d) *Armadillidium vulgare*, 18 mm, (e) *A. vulgare* rolled up, 5 mm diameter

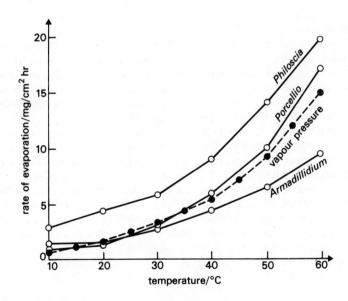

Figure 6.4 The relation between temperature and the rate of transpiration for three species of woodlice. (*After Edney* (*1951*).)

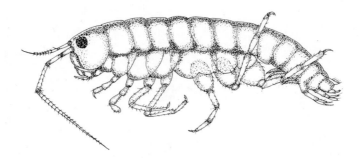

Figure 6.5 A New Zealand species of amphipod recently recorded in Great Britain, *Talitroides dorrieni*, 8 mm

material. On the whole, however, they do not play a significant part in humi-fication.

Amphipod crustaceans are mostly marine but one species of particular interest is *Talitroides dorrieni* (Figure 6.5), common in leaf litter in New Zealand. It was previously described as an introduction from the antipodes and in Europe as an inhabitant of greenhouses. A few years ago it turned up in outdoor localities in Ireland, the Scilly Isles, and, more recently, in Cornwall, being locally abundant in gardens in Penzance where it has been found beneath the Corsican plant, *Helxine solierolii*, and in leaf litter in other parts of western Cornwall.

Amphipods have not achieved the same degree of independence on land as isopods. For one reason, their body shape inhibits progression except by jumping and also prevents them from conserving moisture by rolling up into a ball like *Armadillidium*. Marine amphipods normally feed on algae, thus for their terrestrial counterparts a change of diet to the vegetable matter found in decaying leaves has been necessary and also to living in air rather than water. Copper is important in the diet of all malocostracans and it is thought that in the case of terrestrial forms this is obtained from plant material and made available by micro-organisms. Respiration takes place through the gills, although covering them with paraffin wax does not cause death, which suggests that there are other ways in which oxygen can be absorbed.

Describing the New Zealand talitrids, Hurley (1968) says that they probably evolved directly from supralittoral species being able to enter, with no great difficulty, the environment of leaf litter direct from the supralittoral zone. This would seem to be quite possible for Pacific areas where the forest often reaches the water's edge and where algae and leaf litter are contiguous, but more difficult in western Cornwall and Ireland where there is usually a wide stretch of rocky foreshore. In these areas, individuals which have been successful in reaching land and in establishing colonies there, may have done so in the odd places where ridges of seaweed litter, pushed high up the beach on spring tides, is in close proximity to wooded areas.

Myriapoda

This is an important class of soil Arthropods which can be classified into four groups: the Pauropoda, Symphyla, Chilopoda, and Diplopoda. The first two are part of the permanent subterranean fauna. Less is known about the pauropods, which are seldom more than a few millimetres in length, than about the symphylids, often mistaken for immature centipedes (Figure 6.6). Under

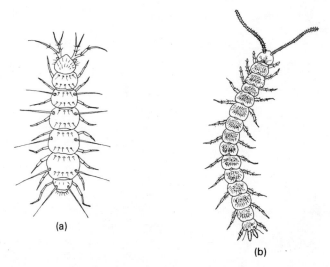

(a)

(b)

Figure 6.6 Two small species of myriapod commonly found in soil: (a) *Pauropus* sp., 1 mm, (Pauropoda) and (b) *Scutigerella immaculata*, 5 mm, (Symphyla)

favourable conditions, greenhouse populations of *Scutigerella* can reach pest proportions feeding on the root systems of cultivated plants. Development from egg to adult in the Symphyla varies with environmental conditions, but takes place over a much shorter time than in most Myriapoda, the larvae usually attaining maturity in about 36 weeks.

Chilopods (centipedes) (Figure 6.7) are mostly carnivorous, they are only partially subterranean being unable to move beneath the surface except by using soil crevices. The well-known red centipede, *Lithobius forficatus*, often found under rotting pieces of wood, has no cuticular waterproofing layer and hence is confined to damp habitats. These centipedes possess powerful jaws and prey on collembolans, mites, small insects, and even other centipedes. The geophilid centipedes, Geophilidae, are also familiar in gardens, woodland litter, and so on. They are capable of forcing their way into the soil. A common species is *Necrophloeophagus longicornis*, which is pale yellow and usually about 30 mm long with up to fifty-seven pairs of legs. Although the cuticle of geophilids has a better developed layer of wax than the lithobiids, they do not seem to be able to withstand desiccation any better, possibly due to the lack of an efficient spiracular closing mechanism.

One other centipede is worthy of mention, not because it is by any means ubiquitous but rather for its astonishing appearance. *Scutigera coleoptrata*,

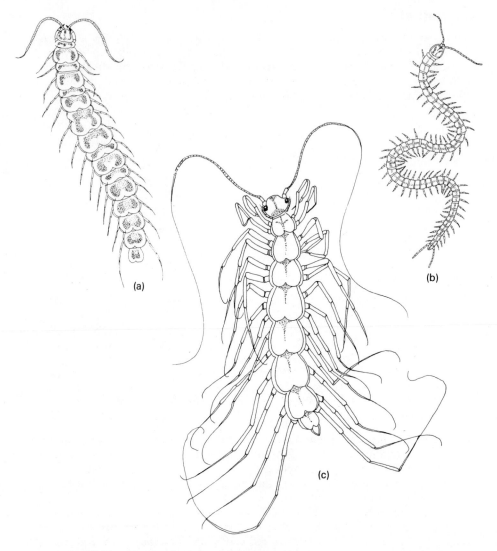

Figure 6.7 Three centipedes (Chilopoda): (a) A common red centipede, *Lithobius forficatus*, 25 mm. (b) A geophilid centipede, *Necrophloephagus longicornis*, 30 mm. (c) *Scutigera coleoptrata*, 25 mm. An introduced species, now well established in the Channel Islands

common in the Mediterranean area of southern Europe and in Africa, is now well established in the Channel Islands and been recorded in places as far apart as Aberdeen, Edinburgh, and Essex where it may have arrived in imported goods.

Scutigera is an altogether handsome animal with fifteen pairs of elongated legs which increase progressively in length from front to back. The whole of the long, annulated tarsus of each leg rests on the surface of the ground as the

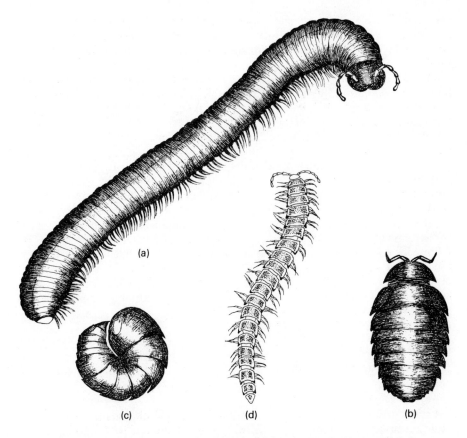

(a)

(c)   (d)   (b)

Figure 6.8   Three millipedes (Diplopoda): (a) Iulid millipede, *Tachypodoiulus niger*, 40 mm approx. (b) Pill millipede, *Glomeris marginata*, 15 mm. (c) *G. marginata* rolled up. (d) Flat-backed millipede, *Polydesmus* sp., 30 mm

animal scuttles about (instead of the clawed tips of the feet, as in other centipedes). The last pair of legs is very long and threadlike, the tarsal portion, like the antennae, being sensory and not used for locomotion. A pair of compound eyes, compared to the simple ocelli of other chilopods, are associated with the swift movements of *Scutigera* as it runs about capturing flies and other insects.

In all centipedes the development from egg to adult takes 3 or 4 years, the total lifespan of *L. forficatus*, for instance, is about 6 years. In *L. forficatus* the larva on hatching has only seven pairs of legs and moults after a few hours. Four more moults take place and at each successive moult the small, pallid larva acquires one more pair of legs until at the final moult it possesses twelve and three pairs of limb buds. Development of this kind is termed *anamorphosis*. After this the larva enters a post-larval epimorphic phase in which there is a slow growth, *epimorphosis*, giving rise to an adult with fifteen pairs of legs. The geophilids hatch from the eggs with a full complement of legs.

The Diplopoda (millipedes), some of which are shown in Figure 6.8, like centipedes, are typical of calcareous, woodland soil, the different species inhabiting various micro-habitats. All millipedes possess a waterproofing cuticular layer which enables them to withstand somewhat drier conditions than most of the Chilopoda.

The pill millipede, *Glomeris marginata*, resembles superficially, the woodlouse, *Armadillidium vulgare* (see Figure 6.3(d)) for both species are capable of rolling up into a ball. The flat-backed millipedes such as species of *Polydesmus*, are mostly to be found in the loose litter layers near the surface. With a less-efficient waterproof cuticle, they are more sensitive to desiccation and cannot therefore survive in places which are either too wet or too dry.

Iulids, because of their cylindrical bodies, can burrow deeper, probably making use of the first trunk segment in their excavations. The glomerids are good burrowers, too. Because of their relatively large size and their habit of feeding on plant litter, attention is naturally drawn to the possibility of millipedes being of importance in the process of humification. Experiments have shown, however, that the amount of litter material converted into humus by millipedes is small. The general picture which emerges is that although they do not make a significant contribution directly to humification, they do break down litter mechanically by their burrowing activities and contribute indirectly as their faeces, constituting relatively large amounts, are more easily decomposed by soil micro-flora than is raw litter.

Annelids

Most soil-inhabiting annelids belong to the Oligochaeta, which includes the ordinary earthworms, of which there are a number of species, and the pot worms or enchytraeids.

Earthworms are not only the most familiar of soil animals but because of their importance in humification processes they have received a great deal of attention which has resulted in a vast literature relating to them. In spite of this, much is still unknown about their ecology, although modern techniques of sampling may make possible a more accurate assessment of their distribution. Their biology and physiology is fully described in numerous books but a note on their reproduction is, however, appropriate.

Among British species, cross fertilization occurs in most members of the genera *Allolobophora* and *Lumbricus* but in other genera such as *Dendrobaena* and *Eisenella*, parthenogenesis is common. (For a key to species see Edwards and Lofty (1972)). Cocoons are laid containing a number of eggs, although usually only one hatches. Conditions being favourable, cocoon production will occur throughout the year, although in different species there may be peak periods which result in the emergence of numerous immature worms.

The distribution of earthworms is largely governed by the pH of the soil (Figure 6.9), most species occurring in neutral or slightly alkaline soils. Earthworm casts can be less alkaline than the soil in which the worms live. One explanation is that the soil is neutralized by intestinal secretions.

Moisture is also a factor influencing distribution, since water constitutes up to 90 per cent of the body weight of an earthworm. Prolonged droughts undoubtedly decrease numbers although worms can take avoiding action in adverse conditions by either moving to moister soil or by aestivating. It has

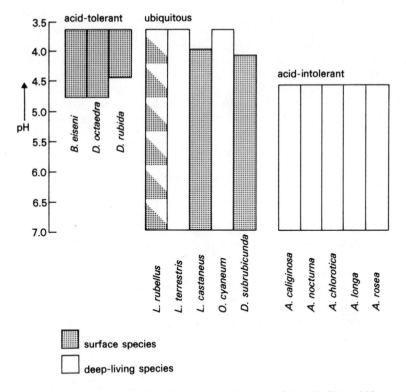

Figure 6.9 Classification of earthworms as a function of the pH of litter. (*After Satchell (1955).*)

been estimated that the common earthworm, *Lumbricus terrestris*, can survive a loss of up to 70 per cent of its total body water content.

Temperature influences the growth, reproduction, and metabolism of earthworms, each species having its own optimum temperature range. For *L. terrestris* this is about 10°C which is lower than for most invertebrates. The emergence of worms at night on to the soil surface is definitely correlated with temperature and the number of leaves buried by *L. terrestris* at different temperatures can be regarded as an index of their activity (Table 6.1).

Table 6.1 The effect of temperature on leaf burial by *Lumbricus terrestris*. (*From Edwards and Lofty (1972).*)

Temperature °C	0	5	10	15
December	0	25	35	67
January	0	33	31	30
February	0	53	63	33
March	0	67	75	44
Total	0	178	204	174

The amount and distribution of organic matter in the soil will affect the numbers of earthworms. Poor soil supports a small population of worms but it is also true that a thick mat of undisturbed organic matter on the surface indicates little activity by earthworms. This can be the case in coniferous woodland. In oak and beech woods where the fallen leaves are palatable to worms, populations are large and they can remove a high proportion of the annual leaf-fall. Soil type also influences both numbers and species. Light and medium loams support a higher total population than heavy clays, alluvial or peaty soils (see Table 6.2).

Table 6.2 Relations of soil type to earthworm populations. (*From Guild (1951)*.)

Soil type	Population thousands/ acre	No./m²	Number of species
Light sandy	232.2	57	10
Gravelly loam	146.8	36	9
Light loam	256.8	63	8
Medium loam	226.1	56	9
Clay	163.8	40	9
Alluvium	179.8	44	9
Peaty acid soil	56.6	14	6
Shallow acid peat	24.6	6	5

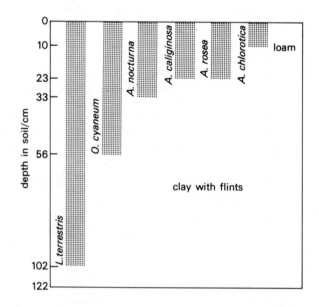

Figure 6.10 Vertical distribution of earthworms in a Rothamsted pasture. (*From Edwards and Lofty (1972), adapted from Satchell (1953)*.)

So far as the vertical distribution of worms is concerned, different species inhabit different zones but also at different times of the year (Figure 6.10). In a field soil *Bimastos eiseni* and *Dendrobaena octaedra* live in the surface organic layers for most of the year while *Allolobophara caliginosa*, *A. chlorotica*, and *A. rosea* are found in the top 8 cm together with immature forms of *A. longa*, *A. nocturna*, and *L. terrestris*. The burrows of mature *A. longa* can penetrate to 45 cm while those of mature *L. terrestris* can go as deep as 2.5 m.

Earthworms make their burrows by literally eating soil and pushing down between cracks and crevices in the soil. Permanent burrows are made by species such as *L. terrestris*, *A. longa*, and *A. nocturna* which burrow deeply. The ejected soil is pressed against the walls of the burrow and bound with mucus secretions which offer a suitable substratum for the growth of certain fungi. It is these burrowing species which produce surface casts near the exits to their burrows, particularly *A. longa* and *A. nocturna*. However, much casting is also done beneath the surface especially in light open soils. The number of casts made is an indication of activity and in autumn a lawn may be covered in casts. Charles Darwin (1881) estimated that in English pastures the production of casts each year was 18.7–40.3 tonnes per ha, or the equivalent of 5 mm in depth of soil deposited annually. It must be remembered, however, that the actual movement of soil accomplished by earthworms is considerably augmented if the voiding of underground casts is taken into consideration.

L. terrestris commonly plugs its burrow with vegetable material such as leaf petioles, leaves and even feathers and stones (Figure 6.11). Food is carefully selected, half-decayed leaves being preferable to newly fallen ones. The leaf blade is eaten leaving the petioles projecting. It may be that the burrow plugs

Figure 6.11 Burrow of *Lumbricus terrestris* plugged with stones and a feather

are to keep water out of the burrow or even to camouflage the entrance. But whatever the reason, if the plugs are removed, the worm will soon re-plug the entrance. In periods of great activity, worms can consume large amounts of food. Guild (1955) found that worms with a body weight of 0.1 g could eat as much as 80 mg of food per gram of body weight per day.

Earthworms are preyed upon not only by birds, but also by badgers, shrews, and especially moles which bite off the anterior segments of the worm and store them in caches in their burrows until needed. A mole requires more than its own weight daily in food, of which worms form the bulk. Carabid and staphylinid beetles and their larvae also prey on worms and so do centipedes and carnivorous slugs of the genus *Testacella* (see below). Earthworms are also parasitized by numerous species of Protozoa, platyhelminths, and nematodes.

Enchytraeidae

Pot-worms, sometimes also called whiteworms, often occur in large numbers where there is an abundance of decaying vegetable matter such as in compost heaps. In the soil their distribution is patchy although in contrast to the other soil annelids, they are more abundant in organic forest and moorland soils than in grassland. Their food consists of plant fragments, fungi, and silica grains, thus they probably make a significant contribution to the production of organo-mineral complexes in the soil. There seems little evidence, however, that they are capable of humification. Unlike the earthworm, they are unable to construct burrows and must use existing crevices in the soil.

Molluscs

The pulmonate gastropods are terrestrial slugs and snails which because of their destructive feeding on cultivated crops, have attracted the attention of agriculturalists more than that of the soil ecologist. This may be because many species really belong to the fauna of above-ground vegetation. However, there are several species of snails, notably the common garden snail, *Helix aspersa*, which lay their eggs in the upper layers of the soil. They are primarily detritus-feeders and so are often found in the soil litter which they ingest. As well as feeding on young plant shoots, some also eat carrion.

Of the twenty-two British species of slug, only three, *Agriolimax reticularis*, *Arion hortensis*, and *Milax budapestensis* are serious pests. *A. reticularis* devours seedlings and soft green shoots on and above the soil surface, while *A. hortensis* and *M. budapestensis* are the most serious pests of root crops. All three species are encountered at different depths beneath the surface, winter frosts and summer droughts causing them to move downwards in the soil to a depth of 22 cm or more. These subterranean forms exhibit a nocturnal activity pattern, emerging at night to forage on the surface. One interesting slug, *Testacella* sp. (Figure 6.12), is a predator of other soil organisms including earthworms. It carries a small ear-shaped shell on the end of the body. The leaf-like pattern of veins enable a rapid change of shape as it pursues its prey through the soil. These it attacks with the needle-sharp teeth of its radula and then works them into its mouth.

The molluscs which feed on surface vegetation and which then move down

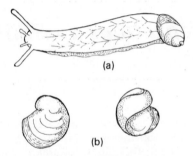

(a)

(b)

Figure 6.12 (a) *Testacella haliotoidea* var. *trigona*. A carnivorous slug, with an external shell, capable of great elongation sometimes reaching 120 mm in length. It will only feed on living animals; earthworms, slugs, snails, and centipedes form its principal diet. (b) Shell, 11 mm, showing growth rings

into the soil are undoubtedly responsible for the incorporation of much organic material. Some produce cellulases in the gut capable of digesting cellulose making it available for other organisms. They also produce copious amounts of mucus which not only assists the soil structure by promoting the formation of soil crumbs, but also provides a substrate for the development of soil micro-flora.

7 Adaptations to Subterranean Life

From previous chapters it will be evident that a large variety of animals live either permanently or temporarily beneath the surface of the soil or amongst the litter on the surface. During their life in the soil they must be able to exist in the absence of light and sometimes in conditions of reduced oxygen supply. Absence of light may involve the development of special sense organs for detecting food or enemies and, in the case of algal-feeding protozoa, restriction to a limited zone near the surface where there is sufficient penetration of light to permit the growth of algae. There is also the question of movement through the soil involving various modifications of body structure for digging, mining or tunnelling.

On the credit side, there may be certain advantages to be had from living in the soil. Temperatures, for instance, do not fluctuate to the same degree as above ground and in the deeper layers of the soil, may be relatively stable. Also, soil can often offer a means of escaping from predators, high air temperatures, drought or even excessive moisture.

Burrowing into the soil may be brought about either by moving the soil mechanically from in front and depositing it somewhere else or by pushing the body through pre-existing soil spaces. Animals that use the former method are usually called 'excavators' and the latter 'tunnellers'.

The excavators

Excavators have literally to dig out the soil either by using their legs or by shovelling forwards with their heads, or by utilizing both methods. Modifications of the legs usually mean that they are flattened and expanded. In the scarabid beetles such as the dor beetle, *Geotrupes vernalis,* there is a slight flattening of the head (Figure 7.1(a)) and of the spines on the tibial joints of the legs (Figure 7.1(b)), which assist with the removal of loosely bound soil. This is done in constructing short burrows for egg-laying or for burrowing beneath dung and litter in the search for food. These modifications are much more extreme in some of the large tropical dung beetles where the tibial spines of the front legs are much more flattened and blunted. The head is expanded and bears a series of rounded spines designed not only for excavation of dung but for rolling along balls of dung, often 5 cm or more in diameter (Figure 7.1(c)).

Many other beetles dig pits in the soil including the larvae of the tiger beetle already described in Chapter 6.

The torpedo-shaped mole, one of the best examples of a fossorial mammal, has forelimbs especially adapted for burrowing with spade-like feet made even broader by the extra sesamoid bone and the flattened claws (Figure 7.2(a)). The mole also uses its snout for forcing its way through the soil. The excavated soil is pushed upwards directly out of the vertical shaft with one front foot, to

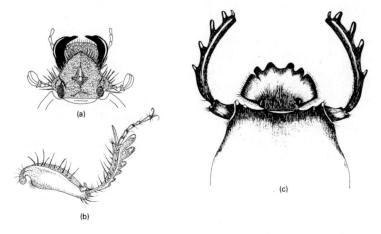

Figure 7.1 Two excavating beetles: (a) Head and (b) foreleg of dor beetle, *Geotrupes vernalis*. (c) Head and forelegs of dung beetle from Ceylon

form a mole hill on the surface. Skoczen (1958) has estimated that as much as 6 kg of soil may be excavated in this way in a matter of 20 minutes. Countrymen maintain that a mole dies immediately if struck on the snout. This may be an indication of the extreme sensitivity of this part of the mole's anatomy which is richly supplied with nerve endings known as Eimer's organs, the importance of which is not fully understood. Nevertheless the very sensitive vibrissae round the snout and its acute sense of smell, undoubtedly make up for the mole's almost total lack of vision. These senses enable it to navigate through the soil and to locate its food which consists mainly of earthworms.

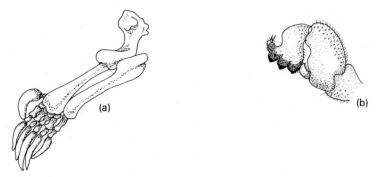

Figure 7.2 (a) Skeleton of left forelimb, 35 mm, of the mole, *Talpa europea*. (b) Left forelimb, 15 mm, of the mole cricket, *Gryllotalpa gryllotalpa*

An excellent example of convergent evolution is afforded by the forelimb of the mole-cricket, *Gryllotalpa gryllotalpa*, which shows a very similar modification of structure to that of the mole (Figure 7.2(b)). The true mole-crickets belong to the orthoptera and although rare in Britain they are common pests of the crops on the continent. They are large insects, about 5 cm long, and excavate galleries in the soil by means of their front legs. They feed on roots and insect larvae. Eggs are laid in the galleries and the nymphs may remain underground for up to 2 years before becoming adult. Even as adults they spend a large part of their time in the soil and so can really be counted as members of the permanent subterranean fauna.

Other orthopteran species which lay their eggs in soil are the great green bush cricket, *Tettigonia viridissima*, and the wart-biter, *Decticus verrucivorus*. The ovipositor of *T. viridissima* is about 20 mm long while that of the wart-biter is sickle-shaped (Figures 7.3(a) and (b)). The females of both species use their ovipositors for excavating the egg burrows in which they lay a single egg moving on to excavate more burrows. The young nymphs, on hatching, climb out of the soil and complete their life cycle above ground.

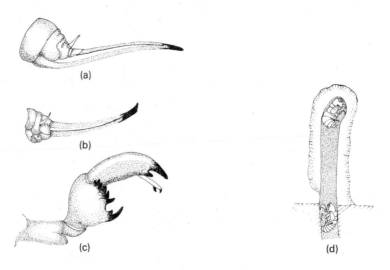

Figure 7.3   Some modifications of insect structure for oviposition and digging: (a) Ovipositor, 20 mm long, of the great green bush cricket, *Tettigonia viridissima*. (b) Ovipositor, 18 mm long, of the wart biter, *Decticus verrucivorus*. (c) Right foreleg, 10 mm, of cicada nymph, *Lyristes plebejus*. The enlarged prehensile forelegs are used as efficient digging tools and also to grip plant roots on which they feed. (d) Longitudinal section through earthen chimney, 40 mm approx. high, constructed above ground for emergence, by cicada nymphs

Among the Hemiptera is one group, the cicadas, which as nymphs are probably one of the best-known fossorial insects. Although only one species is found in Britain there are several on the continent all of which do untold damage to the roots of trees and shrubs. The nymphs have greatly enlarged tibiae (Figure 7.3(c)) by means of which they construct extensive galleries in the soil as they move from one tree root to another, piercing them and sucking the sap. They may spend several years as nymphs underground before rising to the surface to cast the last nymphal skin and emerge as adults. To do this they construct earthen 'chimneys' above ground level up which emergence takes place (Figure 7.3(d)).

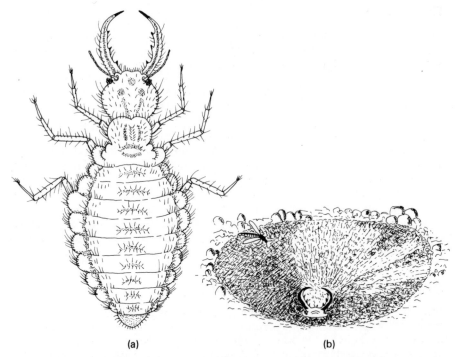

(a)   (b)

Figure 7.4 (a) Larva of ant-lion, *Myrmeleon* sp., 12 mm. (b) Ant-lion larva lying in wait for its prey in the pit it has constructed in sandy soil

One of the most curious creatures which adopts a novel way of catching its prey is the larva of the ant lion, *Myrmeleon* sp. (Figure 7.4(a)). It excavates a circular pit in loose sandy soil by burrowing backwards with its flexible abdomen. Only its head, equipped with powerful jaws, projects above the sand as it waits for some unwary insect to wander too close to the edge of the pit and fall in (Figure 7.4(b)).

Many solitary bees and wasps construct underground nests. Of the solitary wasps perhaps the sand wasp, *Ammophila sabilosa*, is one of the most interesting. The French entomolgist, J. H. Fabre, was one of many naturalists to make minute observations of the activities of *Ammophila* which like other solitary wasps, excavates in loose soil or sand. Working with its mandibles to grapple grains of sand and with the serrated tarsi of its legs, used as highly efficient rakes, it digs a vertical or sloping burrow which ends in a bulb-shaped cell. Several burrows may be made in turn, each provisioned with a single caterpillar first stung into immobility and then dragged overland by *Ammophila* and down the burrow to the cell. A single egg is laid on the inert but living caterpillar and the burrow then carefully closed by grains of sand. In due course the larva hatches and finds the ready-made store of food. Fabre maintained that *Ammophila* was aware of the intimate nerve anatomy of the caterpillar being able to paralyse its prey by stinging each nerve ganglion in turn. Later observers have doubted this somewhat attractive explanation, disclaiming the ability of *Ammophila* to locate the nerve ganglia so accurately.

The tunnellers

Excavating animals depend on removing the soil, while the tunnelling forms rely more upon compressing the soil against the burrow, although some, like earthworms, actually ingest the soil and void it elsewhere.

Animals such as millipedes, wireworms, and pill millipedes have hard integuments and fairly rigid bodies. They can, therefore, literally push their way through the soil or enlarge already existing fissures.

Earthworms are soft-bodied tunnellers which make their burrows by the peristaltic movements of their segmented bodies. For many soft-skinned insect larvae which spend their life in soil, movement is somewhat restricted. The leather-jackets or larval craneflies, (see Figure 6.2(a)) are typical of grassland, feeding on roots. They force their way through the soil by waves of muscular contraction and expansion. The larvae of march flies (*Bibio* sp.) move through the soil in much the same way, assisted by rows of blunt spines in each body segment (see Figure 6.2(c)). The slender geophilomorph centipedes can also progress through loose soil in this way. Many other insect larvae spend their entire existence underground.

Other morphological adaptations

Most of the excavators and tunnellers belong to the meso-and macro-fauna of the soil, but there are modifications of structure among the smaller organisms which enable them to live either in the deeper soil layers or in litter and moss. Most have no specialized means of locomotion and use the water film or soil spaces in which to move about.

The collembolans show an interesting series of modifications connected with the different depths at which they live. Some of the more primitive species such as *Tullbergia* (Figure 7.5(a)) which live in the deeper soil layers, have simple bodies with small legs and antennae, being also blind and colourless and with no springing organ. *Hypogastrura* (Figure 7.5(b)) is an intermediate form, prevalent in litter and moss. It possesses simple eyes, short antennae and a primitive springing organ; while *Orchesella* (Figure 7.5(c)), normally associated with the upper litter layer and often coming to the surface, has longer antennae and legs, also a well-developed springing organ.

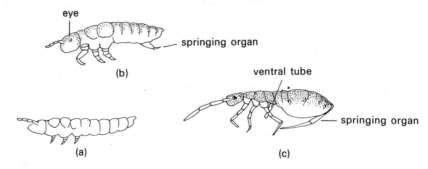

Figure 7.5 Modification of structure in collembolans connected with the depths at which they live: (a) *Tullbergia* sp., (b) *Hypogastrura* sp., and (c) *Orchesella* sp.

Physiological adaptations

Movement of any animal if it is directional, must be governed by its sense organs. In those animals which possess eyes, they are orientated largely by visual stimuli but such stimuli play little part in the lives of soil animals unless it is in the avoidance of light. However, their olfactory and tactile senses may be highly developed, for instance the tactile setae and chemoreceptors of some of the soil mites.

The importance of moisture for soil fauna is evident from the various ways in which they adapt to (for them) either excessive moisture or to the need to conserve moisture. Loss of water might well lead to death by becoming desiccated while too much water in the soil can easily immobilize the smaller animals due to surface tension or the entry into their bodies of excessive amounts of water by endosmosis. Oxygen supplies can also become dangerously low in waterlogged soils. The cuticle of insects is waterproofed by a fine lipoid layer just beneath the epicuticle which endows them with hydrofuge properties preventing the insect from drying up or becoming waterlogged. Other arthropods, notably many of the mites, possess a similar impervious cuticle as well as an efficient tracheal system. A waterproof cuticle does not, however, confer complete immunity to water loss which can also take place during respiration through the tracheae and also via the faeces. This could be serious but, in general, control of excessive water loss is achieved by behavioural responses which cause the mites to seek damper or drier conditions, according to their needs. The intensity of their reactions seems to be influenced by the amount of water contained in the body. Similar reactions are common among woodlice.

Much work has been done on the water relationships of the Myriapoda by Blower (1955) and others which shows that the hydrofuge properties of their cuticles vary considerably. Many millipedes, as well as the geophilomorph centipedes, can survive conditions of excessive moisture better than the lithobioid centipedes, normally confined to the surface layers. The spiracular closing mechanisms in millepedes is more efficient than in centipedes and although the hydrofuge cuticle of both groups prevents water loss, transpiration through the spiracles continues with resultant water loss.

Conditions of drought can be serious, especially for small, thin-skinned animals. Many of the Protozoa form thick-walled cysts. Bdelloid rotifers, tardigrades, and many of the nematodes become immobile, entering a state of anabiosis or suspended animation, at the same time preventing the complete shrinkage of their bodies by the accumulation of oil droplets. With the return of favourable conditions, the cysts dissolve or are cast.

Feeding mechanisms

There is such a variety of methods used by soil animals for feeding according to the type of food they eat, ranging from the plant and fungus feeders to the carnivorous forms which capture their prey in different ways, that there is space only to mention a few of the more interesting adaptations.

The larva of the ant lion, already mentioned, has exceedingly well-developed jaws with which it grasps its prey. Because of its relative immobility, it must rely on the special construction of its burrow and its ability to conceal itself in the sand, to capture small, unsuspecting insects.

The huge chelate pedipalps of false scorpions would seem to be unnecessarily large for the capture of such small creatures as mites and collembola which form their diet. However, close observation shows that once the integument of the prey has been pierced, the mouthparts take over, moving backwards and forwards to macerate the food and add the necessary digestive enzymes. Meanwhile the pedipalps are held in readiness to ward off any attackers. Their trichobothia (tactile hairs) undoubtedly assist in locating prey.

The mouthparts of soil mites are modified according to whether they are phytophagous, saprophagous or carnivorous feeders. As important members of the detritivore chain, there is room for much more work to be done both on their different diets and on their feeding behaviour, not to mention the quantitative aspect of their nutrition generally.

8 Methods of Estimating Populations of Soil Organisms

The soil ecosystem, as we have seen, teems with life and there is also a considerable overlap of species inhabiting the above-ground vegetation, the litter layer, and the soil itself. A qualitative approach to the investigation of soil organisms involves their extraction and identification, but if an attempt is to be made to estimate their numbers, then the methods adopted to extract them from a soil sample, must be those which will ensure that the techniques selected will give the best possible results in terms of the numbers obtained.

Different groups require different methods and for any particular group, none is completely successful for making a total extraction or with every type of soil. Hence the absolute number of animals in a sample is not known and this also means that assessment of the efficiency of an extraction is difficult. That is not to say, however, that comparisons of different techniques cannot be made, and it is the purpose of this chapter to describe those most usually employed. For more exhaustive accounts the reader is referred to Macfadyen (1961 and 1962 a), Murphy (1962), and Edwards and Fletcher (1971).

Estimating soil populations of micro-organisms

Although much valuable information about soil micro-organisms can be gained by direct microscopic examination, this is often difficult in the field although easier for the examination of fungi than bacteria. The soil must be disturbed as little as possible during preparation for its examination. By impregnation with gelatin or with resins (Burges and Nicholas (1961)) thin sections may be made which reveal fungal mycelia and their relationships to various organic substrates such as leaves and the faeces of mites. This method is not applicable to bacteria which are more widely dispersed and thin sections are unlikely to include them. Staining soil particles with phenol aniline blue, however, can show bacterial distribution relative to the particles.

Observational techniques can be used to show changes taking place over a period of time in soil, as well as those described above which indicate the spatial distribution of micro-organisms. The contact slide technique developed many years ago by Rossi involved pressing a clean slide against a freshly cut surface of soil and then fixing and staining the adhering organisms. Cholodney (1930) used the same method but left the slide in contact with the soil over a period of time. Although the technique can provide useful data, it has been criticized on the grounds that the surface of the glass permits the accumulation of moisture and thereby encourages the local growth of bacteria and can even act as a surface over which fungal hyphae not usually present, will grow. Many variations of the Rossi–Cholodney technique have been tried such as the coating of the slide with agar which not only offers a supply of nutrients but encourages the adherance of the organisms to the slides. On balance, however, the method gives a qualitative estimation of the organisms present, quantitative results being possible only if a number of slides are used and this can be of importance in comparing the effect on the soil populations of various nutrients and so forth.

For the estimation of numbers of fungi, bacteria, and actinomycetes in a soil sample, the dilution plate method is probably that most widely used. A known weight of soil is placed in a measured volume of sterile water and shaken thoroughly to ensure the even dispersion of soil particles. Dilutions are then made from the suspension and a measured quantity is poured on to a nutrient medium in a Petri dish. The plate is then incubated and the colonies counted. This sounds comparatively simple but many snags have been encountered. For instance, most soil samples require a dilution of the order of 1:500 000 for bacteria and 1:10 000 for fungi and actinomycetes. Mechanical and not hand shaking of the soil sample is needed, either fairly violently for about 20 minutes or less violently for up to 2 hours, in order to obtain maximum dispersion without damage to the organisms. The use of different media means the selection of certain organisms. Bacteria will grow in a fairly wide range of media so that the choice of the correct one is of extreme importance if specialist groups are to be examined.

Since the range of organisms isolated are influenced by the medium used, to obtain a list of those present in a soil sample, isolations should preferably be made using a number of different media. Even so, there can be wide discrepancies between results obtained by direct methods and by the dilution plate method, which suggests that the latter is only isolating a fraction of the microflora present. Again, bacterial colonies forming on the dilution plate can be regarded as originating from a single bacterial cell, but a fungal mycelium may have arisen from a spore or from a mycelial fragment. Sometimes one species will almost completely occlude the others present, making a count impossible.

Other methods are described by Burges (1958) some of which, especially those involving multiple sampling with at least four replicate plates, have been found to give more reliable counts.

The estimation of protozoan populations in soil is difficult. Although they may be present on Rossi–Cholodney slides, numbers are small and atypical. Normal dilution plates are also unsatisfactory. Singh (1946) pointed out that many protozoans showed specificity with regard to their choice of bacterial food, some species being edible and some inedible. A culture medium inoculated with a soil suspension would promote the growth of a wide range of bacteria resulting in the culture becoming dominated by inedible species. Instead he developed a technique in which plates are poured in a Petri dish from 1.0–1.5 per cent agar with 0.5 per cent sodium chloride. Before the agar sets, eight sterile glass rings are arranged in the dish. After setting, a thick suspension of edible bacteria is spread inside each glass ring. 10 g of the soil sample is shaken for 5 minutes in 50 cm^3 0.5 per cent solution of sodium chloride. Successive dilutions are prepared and the bacterial suspensions within the rings are inoculated with the different dilutions and incubated at 22°C. Drops of sterile 0.5 per cent sodium chloride are added to the cultures to prevent them from drying out. Samples are taken from each of the rings and the number of protozoa per g calculated using Fisher's method (see Singh, ibid).

It will be evident from the foregoing paragraphs that no infallible technique

Table 8.1 Relative number of organisms per gram in a fertile agricultural soil.

Organisms		Number of organisms
Bacteria	Direct count	2 500 000 000
	Dilution plate	15 000 000
Actinomycetes		700 000
Fungi		400 000
Algae		50 000
Protozoa		30 000

has been devised for the estimation of numbers of the different groups of micro-organisms. Comparisons of the various methods employed only emphasizes the diversity of results. The number of organisms recorded for different soils varies enormously. For instance Table 8.1 merely shows the *relative* numbers of organisms present in a fertile agricultural soil, the numbers of bacteria varying the most. In a soil with pH greater than 6.0 there can be counts in excess of 10 million per g pH does not affect fungal counts so much. It must always be remembered, however, that any soil will show a fluctuation in numbers of the different groups and species not only influenced by pH but by biotic factors and by changes in temperature and humidity.

Techniques of laboratory extraction of animals other than micro-organisms

The principal techniques can be described under two headings: those by which the animals are physically separated from the soil by mechanical methods and those in which the animals are driven from the soil by employing the behavioural responses of animals to stimuli such as heat, illumination, and desiccation.

Of the mechanical methods, sieving can be used in the case of animals larger than the particles of soil, for instance earthworms. For some animals, particularly enchytraeids, dry sieving is not satisfactory and a method of wet-sieving must be used (Figure 8.1). The litter sample is supported on coarse mesh and continuously stirred so that the worms become detached and collect on a fine cloth covering the lower grid.

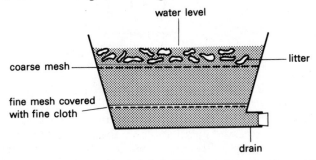

Figure 8.1 Wet-sieving apparatus for the extraction of enchytraeids from leaf litter

Flotation methods are employed for some groups of micro-arthropods, such as mites, which cannot be so easily extracted by sieving because of the large amounts of plant debris involved. The 'Ladell' technique, which uses quite complicated apparatus, is fully described by Wallwork (1970). This method involves first the separation of mineral soil from organic matter followed by the extraction of the micro-arthropods from the plant material by differential wetting.

Sieving and flotation methods, although widely used and capable of the simultaneous extraction of a variety of animal groups, have their disadvantages. Such methods cannot be used for soils with a high organic content, because they extract both living and dead organisms without distinction and, most important, are time-consuming. For these reasons many soil zoologists prefer to use behavioural techniques.

The Berlese-Tullgren funnel method and its various modifications are probably those most generally used in dry-funnel extractions. Basically the apparatus consists of a steep-sided, smooth-walled funnel, often made of glass. In the mouth of the funnel the soil sample rests on gauze over which is suspended a carbon-filament lamp to provide a source of heat and light. As the temperature of the soil sample rises the animals move downwards and away from the source of heat, falling into a collecting tube. Figure 8.2 shows Mac-

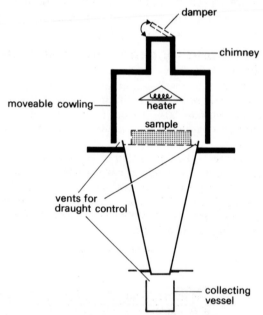

Figure 8.2   Macfadyen's controlled draft funnel

fadyen's modification in which the circulation of air is controlled. Points to remember are that for successful operation there should be steep gradients of both temperature and humidity. It is important, therefore, that the soil specimen should not completely cover the gauze but should allow for a passage of air thus also reducing the amount of condensation on the sides of the funnel.

Placing water in the collecting vessel instead of alcohol will increase the moisture gradient. In the controlled draught funnel, extraction begins with the use of a 10 W bulb and the lid opened to 90 degrees. The surface of the sample gradually warms up but no extraction, in this first stage, takes place. The bulb is exchanged for one of 15 W and the lid shut but should this be done too soon, condensation may result. With the sudden increase in temperature, animals begin to emerge. The lower part of the sample remains moist and there is a reduction in activity followed by a secondary emigration as the lower layers of the soil sample become desiccated. If extraction is to be made from soil cores, as distinct from litter, it is advisable to place the core undisturbed, but inverted on the gauze, thus permitting the animals near the surface to escape without burrowing. Breaking-up of the core can result in damage to smaller soft-bodied species which, seeking refuge in damp crumbs of soil, can become incarcerated as these dry out. The sides of the funnel should be dry and polished at the start. A frequent cause of trouble is the trapping of organisms in moisture on the sides of the funnel (hence the use of one which is steep-sided) or in webs spun on the sides by small spiders.

Some workers have abandoned the use of large funnels, the tendency being nowadays to employ many smaller funnels thus allowing for an adequate number of replicate samples, or even to collect the animals directly into tubes of the same diameter as the sieves supporting the soil.

In wet-funnel methods, such as Baermann funnels (Figure 8.3), the sample can be contained in a mesh bag of some kind suspended in the mouth of the funnel which is completely filled with water. A source of heat above causes a

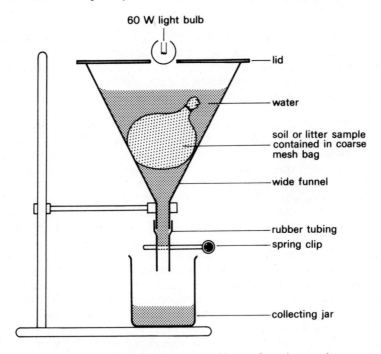

60 W light bulb

lid

water

soil or litter sample contained in coarse mesh bag

wide funnel

rubber tubing

spring clip

collecting jar

Figure 8.3 Baermann funnel used in wet-funnel extraction

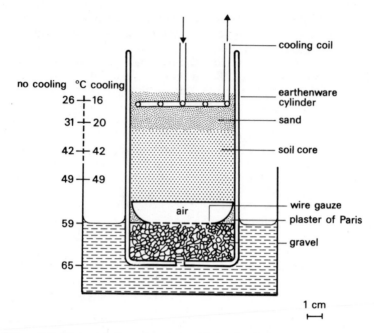

no cooling °C cooling

no cooling	°C cooling
26	16
31	20
42	42
49	49
59	
65	

cooling coil

earthenware cylinder

sand

soil core

wire gauze

air

plaster of Paris

gravel

1 cm

Figure 8.4 Wet extraction of enchytraeids by the Nielson extractor

temperature gradient in the water and the organisms to pass downwards, through the mesh container into the water. Eventually they go down the stem of the funnel, fitted with rubber tubing and a spring clip, and can be collected in a tube. Wet-funnel methods are particularly useful for the extraction of animals normally associated with the soil water, such as nematodes, rotifers, and enchytraeids.

Another method of extracting enchytraeids has been used by Nielson (1953). The apparatus (Figure 8.4) consists of a cylinder in the bottom of which is a layer of gravel. The cylinder is immersed in a water bath so that the level of the water in the bath corresponds to the level of the gravel. The soil sample is held above the wire gauze by a platform of plaster of Paris and covered with a layer of sand in which there is a cooling coil. The water in the bath is heated at a rate sufficient to produce a temperature of about 60°C in approximately 2 hours, this establishes a temperature gradient of about 60°C to 30°C throughout the sample. At the same time water evaporates from the gravel and moves through the soil sample where it condenses because of the lower temperature. Thus a moisture gradient is also established. Upward migration of the enchytraeids begins at once and after about 2 hours they will have accumulated in the upper part of the sample. In order to concentrate them in the sand, the water bath is heated to 85°C and cold water is circulated through the coil. This produces an abrupt decrease of temperature in the sand. Extraction can be stopped after 3 hours and the enchytraeids separated from the sand by flotation.

A comparison of laboratory extraction methods

In productivity studies it may be necessary to assess as accurately as possible, populations of soil organisms at different sites and in the comparison of the productivity of ecosystems (see Chapter 10) it would be an advantage to be able to standardize methods. However, working conditions, soils, and their fauna, not to mention the human element, differ.

No single method is best for all groups of soil animals and it is advisable to employ several sample sizes and several methods of extraction in order to obtain some assessment of a population.

The reader is referred to Edwards and Fletcher (1971) for comparisons made by different methods of funnel extraction. Their work showed that there was a wide variation in numbers of organisms extracted by the different methods, the Macfadyen funnels giving the most complete extractions.

In general, funnel extractions are better suited to highly organic soils or litter while flotation methods, although more time-consuming, are more efficient for mineral soils.

Field sampling

Collecting animals in the field requires different methods. For surface-dwelling insects, isopods, and spiders various forms of pitfall traps can be used, baited with decaying foods such as fish, meat, and melon. Unbaited traps are usually visited by carnivorous species; but as a means of estimating population densities, such traps are of little value since they may attract organisms from places remote to the actual area investigated. The reaction of a species to trapping, may vary with its age, diurnal and seasonal activities, and with its reproductive cycle. Hence, results may be more an indication of the activity of a population than of its size.

The application of certain chemicals can be used for extracting earthworms and geophilomorphid centipedes. Potassium manganate (VII) (potassium permanganate) solution was used by Evans and Guild (1947) in their pioneering experiments on earthworm extraction but the method has been shown to give considerable underestimates of population density. Raw (1959) used methanal (formaldehyde) to expel earthworms and this has been found useful for studies of *Lumbricus terrestris* and certain other species including *L. castaneus*, *Allelobophora chlorotica*, *A. longa*, *A. rosea*, and *Octolasium cyaneum*. Raw tested a solution of 25 cm³ of 40 per cent methanal in 4.55 dm³ which is approximately a 0.55 per cent solution. The rate of application was 12 dm³/m² applied in two lots, the second about 20 minutes after the first. For greater accuracy Satchell (1971) recommends that three lots of 9dm³ of 0.275 per cent solution should be applied at 10 minute intervals to each quadrat of 0.5 m², using a number of samples. Table 8.2 gives his results from which it is seen that the more active and/or surface-dwelling species such as *L. castaneus*, tend to emerge sooner than the deeper-burrowing forms. Temperature and soil moisture both have an effect on the number of worms expelled and Satchell found that the maximum response by *L. terrestris* to methanal occurred at a temperature of 10.6°C and the numbers expelled increased linearly with soil moisture content to 38 per cent decreasing at 44 per cent. For correction of temperature and soil moisture the appropriate model to use (Lakhani and Satchell (1970)) is:

Estimated population=
Observed population × exp (0.0075 $(T-10.6)^2$)×exp (-0.0214 ($M-40$))
where T =soil temperature at 10 cm in °C
and M =percentage soil moisture content.

Table 8.2 The rate of expulsion of earthworms using successive applications of methanal (formaldehyde) solution at 10 minute intervals. (*From Satchell (1971)*.)

	Number of worms (%) expelled between each application							Total number expelled
Application	1	2	3	4	5	6	7	
L. castaneus	79	19	1	0	1	0	1	85
A. chlorotica	74	18	3	2	1	1	0	407
L. terrestris	63	29	5	2	1	0	0	318
A. rosea	51	34	7	4	2	1	0	268
O. cyaneum	47	27	15	4	0	2	5	55
A. longa	41	41	10	0	3	3	3	39
A. caliginosa	40	37	16	5	1	1	0	141
All species	61	27	7		5			1313

Because terrestrial molluscs are important crop pests various field methods have been tried for estimating populations. Numbers of surface-dwelling snails, such as *Cepaea nemoralis* or *Helix aspersa* are relatively easy to estimate since they can often be counted by collecting them either from within randomly distributed quadrats or along a transect.

Mark and release methods have also been used with snails since they can be marked on their shells and provided the animals mix randomly with the unmarked population and that sufficient numbers are recovered (greater than 10 per cent of the total released), this system is quite satisfactory.

Slugs are less easy to mark although some workers have labelled slugs with radio-isotopes which they can be made to ingest, so that they can be traced underground, which has the advantage that the site is, left relatively undisturbed. Indirect methods include the use of slug traps using bait such as ethanal tetramer (metaldehyde) or bran. Results, however, depend on the average night temperature and also upon the length of time the bait has been put down. Webley (1964) found that the higher the temperature the greater the numbers of slugs caught and therefore baiting does not give reliable results.

9 The Soil Community at Work

The organisms living in the soil, some of which have been described in previous chapters, form a community all adapted in one way or another to living in the soil environment and therefore they are a part of the soil ecosystem. This chapter describes the special environment of roots and how the functioning of the soil community is concerned with the decomposition of organic matter and the release of inorganic compounds making them available to higher plants growing in the soil. Although the soil community is here treated as a unit, in reality there are many communities within the soil, all intimately connected. There are, as well, obvious overlaps between populations inhabiting the soil with those living above the surface.

The environment of roots

Living plant roots provide a source of nutrients in the form of exudates, as well as from their living tissues, for micro-organisms which occur in greatly increased numbers and show an increase in activity in proximity to roots. This is called the *rhizosphere* which, although it has no real boundary, is the area of soil round the roots which is modified by the roots. Some organisms of the rhizosphere form associations with the root surface, others actually invade the living root cells where they live as symbionts or parasites.

Without doubt the metabolic activity of roots affects the environment around them by the excretion of carbon dioxide thereby altering the pH of the surrounding soil, and in turn causing an increase in the activity of many micro-organisms. Apart from this, the surface cells of roots are continuously being sloughed and in young roots the surface can often be covered with mucilage which provides a source of nutrients for micro-flora.

The microbial populations surrounding roots are influenced by many factors including the kind of plant, its age and vigour, the type of soil, and the environment. A peak of rhizosphere effect is reached around the time of fruiting, declining as root senescence begins. The effect of the rhizosphere is seen at its best in sandy soils, sand-dune plants causing a marked increase in the micro-flora of the surrounding soil.

Reference has already been made to the close association of some of the micro-flora with plant roots. Symbiotic associations involving fungi and plant roots are common and are known as mycorrhiza. Most of our forest trees possess ectotrophic mycorrhiza, the fungus forming an external felted covering to the roots, only penetrating the epidermal cells. Infection takes place through the root hairs of the seedling tree and although it spreads rapidly, the fungus does not seem to be essential to the tree's existence, it is probable that the mycorrhizal fungus makes inorganic nutrients available to the tree. The fungi involved are mainly basidiomycetes – species of *Amanita*, *Boletus*, and

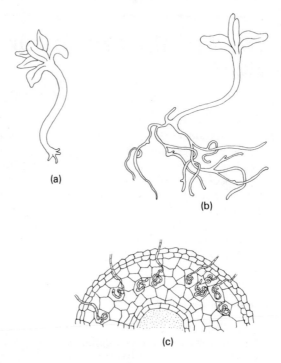

Figure 9.1 Seedling of heather, *Calluna vulgaris*, 5 months old: (a) grown in sterile culture and (b) infected with mycorhizal fungus. (c) Transverse section of orchid root showing penetrating fungal hyphae

Tricholoma – known by their large fruiting bodies.

Endotrophic mycorrhiza occur in all the orchids and all the heaths. The relationship is not fully understood although in both groups the association is necessary for the proper development of the seedling plant (Figure 9.1).

Infection takes place from the soil, the penetrating hyphae entering the root cells to become coiled within the cells as they grow.

The circulation of plant nutrients in the soil

The principal elements required by plants are hydrogen and oxygen, both of which are present in the environment in large quantities. Carbon and nitrogen, however, and to a lesser extent phosphorus and sulphur, are also essential to plants and would soon become deficient in soil unless they were recycled.

The carbon cycle

Green plants obtain their carbon from the carbon dioxide in the atmosphere and during the process of photosynthesis are able to fix the carbon, converting it into sugar. This brief statement, however, is very inadequate for a process which, in reality, is highly complex and involves many stages. Carbon dioxide is present in the air at a concentration of 300 ppm which may be less in the vicinity of actively photosynthesizing plants. Some carbon is returned to the atmosphere by the green plants themselves during respiration, some is excreted by their roots but most is incorporated into the plant tissues as cellulose, lignin or as the compounds forming protoplasm and storage materials. All the carbon stored in this way is eventually returned to the soil when the plants die and their bodies decompose. Some idea of the quantity of carbon returned in this way can be gained from figures calculated for a tropical rain forest where both growth and decay are greatly accelerated. Here something in the region of 9000 kg of carbon per hectare are returned to the soil as dead branches, leaves, and roots.

Plants can, however, be eaten by herbivores which, in turn, are devoured by carnivores so that the carbon fixed by a green plant may pass through several animals before becoming part of the dead organic matter in the soil.

A glance at Figure 9.2 will show from what sources this reservoir of dead matter in the soil emanates. In order to reach a state of equilibrium the rate at which decomposition takes place must be fast enough to remove this organic matter, without it accumulating, and thereby to replenish the supply of atmospheric carbon dioxide. The processes by which this reduction takes place

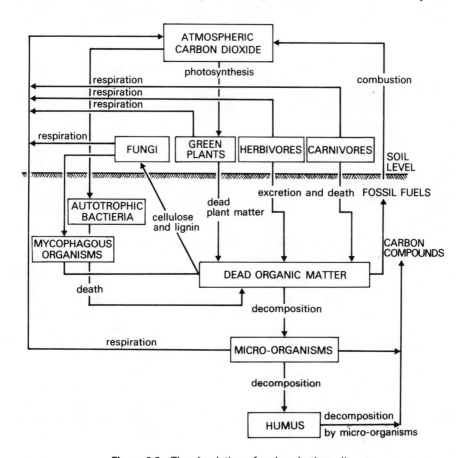

Figure 9.2 The circulation of carbon in the soil

are complex, passing through a number of stages in which the more readily digestible carbohydrates, lipids, and proteins of the plant body are first decomposed followed by the more slowly decomposable materials such as lignin and cellulose. In forest soils the latter are largely decomposed by fungi which are themselves consumed by mycophagous animals such as mites, collembolans, and nematodes.

Some of the complex organic compounds synthesized by micro-organisms in the first stages of the decomposition of organic matter, are resistant to digestion by other organisms and may accumulate as humus which is always slowly decomposing.

In addition to the organisms which bring about the decomposition processes there are some autotrophic bacteria capable of synthesizing carbon dioxide directly from the atmosphere.

At all stages of this cycle, the organisms taking part, whether of plant or animal origin, are releasing some carbon dioxide during their respiratory processes.

Finally, one important source of carbon dioxide is the combustion of carbon-containing solid fuel which releases carbon formed as the result of photo-synthesis by plants which have become fossilized many millions of years ago.

The nitrogen cycle (Figure 9.3)

Although plants live in an atmosphere containing much more nitrogen than carbon, it is in a form which they are unable to assimilate directly from the air. Nitrogen is only usable by most plants in the form of nitrate compounds and the green plant is dependent upon a chain of bacterial reactions to make these compounds available.

One group of nitrifying bacteria of great importance to the soil ecosystem consists of those symbiotic in the roots of leguminous plants, which are capable of direct fixation of atmospheric nitrogen. These bacteria belong to the genus *Rhizobium* and enter a root through the root hairs. Here they multiply and travel to the cortical cells causing hypertrophy of the root cortex. The nodules so formed eventually disintegrate and the bacteria are liberated in the soil. If a suitable host plant is present the rhizobia are stimulated and large populations occur in the rhizosphere. In these regions concentrations can be up to as many as 10^7 individual cells per gram of soil.

Other important but free-living aerobic nitrogen-fixing bacteria are *Azobacter*, *Clostridium*, and *Pseudomonas*, which can convert atmospheric nitrogen to ammonium compounds. Two autotrophic nitrifying bacteria are responsible for oxidizing these ammonium compounds to nitrate (V) compounds. The first, *Nitrosomonas europea*, converts these compounds to nitrate (III) compounds and *Nitrobacter agilis* completes the process by converting nitrate (III) to nitrate (V) compounds. The former are phytotoxic but the second reaction occurs very rapidly thus preventing a build-up to toxic proportions. Although these nitrifying bacteria grow rather slowly, large quantities of ammonium compounds can be converted to nitrate (V) compounds, one estimate made showed that between 7.9 kg and 98.7 kg per hectare per day could be formed in loam and clay soils which had been treated with ammonium fertilizers.

As with the circulation of carbon, the breakdown of dead organic matter and

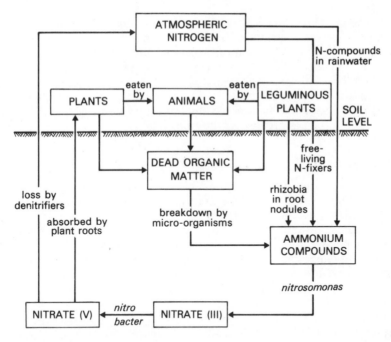

Figure 9.3 The circulation of nitrogen in the soil

its subsequent build-up into nitrate (V) compounds by bacteria, plays an important part in the recycling of nitrogen. Some of the salts so formed are lost by leaching which carries them down into the soil below the depth at which they can be absorbed by plant roots. Some is also lost by denitrifying bacteria such as *Pseudomonas denitrificans* as well as by some denitrifying fungi which liberate nitrogen from nitrate (V) compounds especially in poorly aerated soils with a high organic content. Improved soil aeration by ploughing and draining assists against such losses.

The sulphur cycle
Sulphur, also an essential constituent of plants, accumulates in the soil as organic compounds which are not available to plants until they have become converted to soluble sulphates by the action of micro-organisms. Like the denitrifying bacteria there are those which can reduce sulphate under anaerobic conditions with the release of hydrogen sulphide.

The phosphorus cycle
Several important compounds, such as the nucleic acids (nucleoproteins of which phosphorus is an essential constituent), are present in the bodies plants and animals. In the soil phosphorus occurs as both mineral salts and as organic compounds. Many plants absorb their requirements of phosphorus as orthophosphate. Plant and animal residues give rise to organic compounds of phosphorus and these are also used by micro-organisms which therefore compete for the available phosphorus. However, certain bacteria and fungi can convert the phosphorus present into orthophosphate chiefly by means of

organic acids. This process is particularly active in the region of plant roots which produce sugar exudates. It is also possible that plants with mycorrhiza may be able to utilize phosphorus otherwise unavailable to them due to the presence of their fungal partners. Curiously enough, much of the phosphate-containing fertilizers such as superphosphate, added to the soil with the intention of increasing the supply of phosphorus, can be rapidly converted by chemical fixation to inorganic compounds, some being thereby rendered unavailable to plants because of their insolubility.

Humification

It is apparent from the previous sections that many of the organic processes affecting the life of plants and soil animals takes place by soil micro-organisms beneath the surface of the soil.

When one considers that plant residues may range from succulent leaves to highly lignified woody remains and that animal residues include complex organic compounds ranging from bone to partially digested foods, it is not surprising that an army of micro-organisms, each responsible for an organic or chemical change, is required to bring about decomposition. Both chemical and physical changes are involved.

The processes contributing to the production of substances such as carbon dioxide, water, nitrites, nitrates, sulphates, and phosphates are referred to as *mineralization* and those which lead to the formation of organic residues, often resistant to change, as *humification*.

Most raw humus is deposited on the surface as leaf litter or the dead remains of animals and plants, but there is a by no means negligible contribution made underground by the exudates of root systems as well as by faunal residues. This fresh organic matter can be decomposed or mineralized by the microbial flora of the soil involving bacteria, actinomycetes, and fungi as well as some algae and Protozoa. The bacteria and actinomycetes are mostly heterotrophic, utilizing as food organic carbon and nitrogen compounds. Some are autotrophic, synthesizing food from simple inorganic substances. The fungi are probably all heterotrophic while the algae, because they possess chlorophyll, can make their own carbohydrates. The Protozoa mainly feed on bacteria.

A quantity of fresh organic matter is incorporated directly into the soil by earthworms which consume plant debris and mineral matter, partly digesting the organic fraction and excreting both. The part played by earthworms in the first stages of humification is very considerable due to the enzymatic digestion of organic material as it passes through the gut assisted by the gut flora which decompose lignin and cellulose. The same probably applies to the soil micro-arthropods.

The general picture presented in the decomposition processes is as follows: invasion of the organic debris by the first decomposers is usually rapid, the utilizable materials being converted into the body material of the decomposers with the evolution of carbon dioxide and the breakdown of certain minerals which are reabsorbed by plant roots. The residual material that is left is then attacked by a second wave of decomposers. Once more, synthesized material is incorporated in their bodies with the evolution of carbon dioxide and the release of more mineral matter. There is now additional residual material

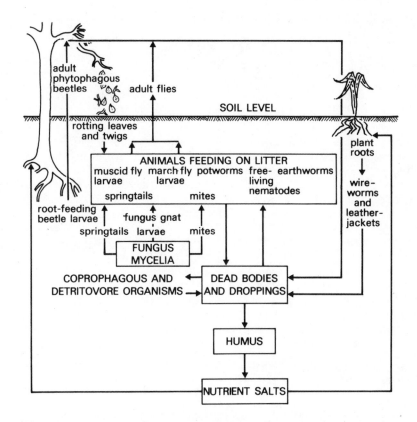

Figure 9.4 Schematic diagram showing some food relationships of soil organisms and the part they play in the process of humification

formed in the bodies of the first decomposers and also that left by them. As each succeeding wave of decomposers break down the remaining debris, the available material decreases until it is finally incorporated into the mineral soil as humus.

Our knowledge of the role of Collembola, mites, and other small organisms in the formation of humus has at present not advanced much beyond the study of populations. Investigation concerning their fundamental biology and the part they play in the breakdown of organic material must progress further before the schematic diagram in Figure 9.4 can be accepted as anything more than an estimation of food relationships and the processes involved in humification. There is little doubt, however, that with the exception of earthworms, the contribution made by the meso-fauna to the process of humification is slight, although their burrowing and tunnelling activities result in the physical break-up of the soil and in its better aeration and drainage. Millipedes assist in this way but may also be important in the chemical breakdown of litter in such a way that other agencies can complete the process. However, the method and extent of this breakdown is still not fully understood.

10 Dynamic Aspects of the Soil Community

The major source of energy in any terrestrial ecosystem is from the Sun. This solar energy is converted by plants into chemical energy during the process of photosynthesis and used by them for growth and respiration. Respiration involves the loss of energy in the form of heat which is therefore not available to other organisms. Some of the energy stored by plants in their tissues is removed by being eaten by grazing animals. The rest reaches the soil in the form of dead plant material such as leaves, woody stems, and roots. Soil herbivores, which include smaller organisms as well as the soil micro-flora, feed on these plant residues.

It would be convenient to be able to determine two pathways of energy flow – one the 'grazing chain' and the other the 'detritus chain' – as is often possible for other ecosystems. However, in the soil complex the two chains are not clearly identifiable since many soil animals forming part of the grazing chain feed on a variety of organisms belonging to both chains and themselves provide food for a heterogeneous collection of organisms.

The regulation of natural populations

Although not in itself directly connected with adaptation to life in the soil, the ways in which populations are regulated may involve many adaptive mechanisms designed to maintain numbers. The tendency of populations to decrease due to death or emigration is offset by the tendency to increase due to birth and immigration. A stable population, in itself a rare phenomenon, is one in which these two opposing tendencies are equated. It is more likely, however, that their interaction results in regular and often quite sudden fluctuations. In theory this sounds a reasonable premiss which is difficult to assess in practice due to lack of adequate information. Much more work still needs to be done on food relationships generally especially among the meso-fauna, about which we know relatively little.

It is true to say that no one form of regulating mechanism can be applied to communities comprising different species. Predation is probably the prime determinant in regulating the numbers of Collembola and Nematoda which form important links in many food chains. Another limiting factor can be temperature, which affects the length of the reproductive periods. Fecundity can also be greatly reduced when food is short, especially in animals such as earthworms which have limited food preferences and which select material with a high nitrogen content. This, in itself, can lead to competition, another important regulatory factor.

Table 10.1 The numbers, weights, and oxygen consumption of various groups of animals in Danish forest soils. (*After Bornebusch (1930).*)

Animals	Mull soil under beech (pH 6.1–5.8)			Raw humus under beech (pH 5.6–3.6)		
	Numbers in millions per hectare	Percentage contribution to the total weight of animals present	Percentage contribution to the total oxygen consumption of animals present	Numbers in millions per hectare	Percentage contribution to the total weight of animals present	Percentage contribution to the total oxygen consumption of animals present
Earthworms	1.78	75.1	56.2	0.81	22.4	12.2
Enchytraeid worms	5.34	1.5	6.0	7.81	6.5	14.4
Gastropods	1.04	7.0	7.3	0.52	13.4	7.5
Millipedes	1.78	10.6	15.0	0.40	4.7	3.9
Centipedes	0.8	1.8	3.5	0.2	2.1	2.0
Mites and springtails	44.12	0.4	4.5	112.46	2.3	16.1
Diptera and elaterid larvae	2.45	2.4	3.7	13.05	43.8	35.2
Other insects, isopods, and spiders	4.72	1.2	3.8	6.5	4.7	8.7

	Mull soil under beech	Raw humus under beech
Total number of animals	61.79 million per hectare	141.37 million per hectare
Total weight of animals	286 kg	97 kg
Oxygen used per m² at 13°C	0.33 dm³ per day or 90 g per m² per year	0.20 dm³ per day or 60 g per m² per year

(i.e. 1 g of O_2 is needed to oxidize 1 g of organic matter to CO_2)

Population metabolism

The highly organic soils of forest vegetation can support a varied fauna typical of either mull or mor and are the result of the interaction of several factors of which the species of tree, the ground vegetation, and the parent material of the soil all play a part. An early survey of the soil fauna of woodland sites carried out by Bornebusch (1930) showed that the total weights of each animal species did not correspond to the quantity of organic matter oxidized by the animals, for the smaller species required more oxygen for respiration. Table 10.1 gives a summary of Bornebusch's results of the live weight and oxygen requirements of each group of animals in two typical soils under beech forest. The mull carried a flourishing ground vegetation while the mor (raw humus) was almost devoid of vegetation.

Mites and spring tails, so far as total numbers were concerned, formed almost the entire population although their total weight was almost negligible. Their contribution to the total oxygen demand was, however, appreciable.

Table 10.1 also shows that the oxygen demand of soil fauna does not necessarily increase with an increase in the total number of animals present. Later investigators have shown that Bornebusch's estimates of density and biomass are too low although the general conclusions as to the relative distributions of the more important groups of soil organisms are valid.

On two forest soils on which Bornebusch sampled earthworms, he found that 2.5–3.75 million worms per hectare weighed 1700–3000 kg approximately, which equalled the weight of livestock carried per hectare on first-class Danish pastures. However, the amount of carbon dioxide respired by the earthworms was only about one-tenth that of the livestock.

Community structure

Broadly speaking a biological community is an integrated assemblage of plants and animals living together in an environment. But in attempting to compile a scheme of classification, at different levels of complexity, one immediately discovers that many soil communities are transitional, one merging into another. For instance, many of the organisms associated with litter also play an important part in the soil itself. If however, we accept that the species is the basic ecological unit, true communities are those composed of a number of species usually associated in a characteristic way. It is most important, however, that a community should be considered as a dynamic system within which there is a flow of energy between the species.

Certain concepts involved in the flow of energy must be defined. For instance the terms 'standing crop' and 'production' are often confused. The *standing crop* of a population of living organisms is a measure of the numbers or quantity of those organisms at one point in time, or the average of these measures at different points in time. Such measurements are usually expressed in terms of numbers or weight per unit area.

Table 10.2, adapted from Nef (1957), in which there are certain important gaps, compares the numbers, weights, and respiratory activity of invertebrates in two types of forest soil. Such a table, although useful in recording and comparing numbers, tells us little of the dynamic aspects of these communities.

Table 10.2 Approximate estimates of numbers of invertebrates in two types of forest soil habitat, indicating biomass and respiratory activity. (From Kevan (1962), compiled from various authors and adapted from Nef (1957).)

Group	Spruce plantations			Deciduous forest with mull		
	Thousands per m²	Biomass g/m²	Respiration at 13°C mg O₂/hr/m²	Thousands per m²	Biomass g/m²	Respiration at 13°C mg O₂/hr/m²
Protozoa	100 000	2	?	200 000	4	?
Turbellaria, etc.	?	?	?	?	?	?
Rotatoria	200	0.1	?	600	0.3	?
Nematoda	1700	4.5	5	20 000	10	10
Mollusca	—	10	—	0.1	5	1.5
Enchytraeidae	75	10	10	30	4	4
Lumbricidae	0.03	1.5	0.59	0.12	60	11
Tardigrada	25	0.05	?	100	0.2	?
Crustacea	—	—	—	0.28	0.3	0.4
Arachnida (except Acarina)	0.067	0.07	0.09	0.06	0.06	0.08
Acarina	400	4	12	400	4	12
Diplopoda	—	—	—	0.11	4.7	1.85
Chilopoda	0.12	1.8	1	0.04	0.6	0.3
Protura, Diplura, etc.	?	?	?	?	?	?
Collembola	100	1	3	200	2	6
Insecta (saprophagous)	0.8	5.3	2.9	0.4	4	1.5
Insecta (carnivorous)	0.25	0.5	0.56	0.08	0.7	0.4
Total Metazoa	>2500	28.82	35.14	>21 330	95.86	49.03

The concept of the production or productivity of a community, however, is essentially a dynamic one for it is a measure of the amount of material and energy stored by an organism, or population of organisms, within a certain period of time, and is usually expressed as joules per unit area per unit time. The *gross production* of a community is the total amount of energy stored by the organisms concerned. *Net production*, on the other hand, is the actual amount of energy used in growth. The difference between gross and net production is that used up in metabolism and normally expressed as a measure of the respiration of the species involved.

Food chains and food webs

All ecosystems have certain features in common. Within each system the numbers of organisms present depend upon the amount of energy available and this is closely linked with the way in which energy is transferred from one feeding grade to another through the different trophic levels. As already mentioned in Chapter 3, the chief classes of organisms concerned with energy transfer are:

(1) the *primary producers* or those capable of utilizing the Sun's energy directly, synthesizing their own food in the process of photosynthesis. Primary producers are the higher plants, certain bacteria, and most algae.

(2) The *primary consumers* or those animals which feed directly on plant material – the herbivores.

(3) The *secondary consumers* which feed only on animal material – the carnivores. In practice there are some organisms which feed on both plant and animal matter (omnivores) and also parasites which feed on organisms at all three levels.

(4) The *decomposers* are an important group of organisms which, again, may belong to any of the first three groups but which have the special function within a community of breaking down dead or decaying matter into simpler substances with the release of inorganic salts, making them available once more to the primary producers. Chief among these decomposers are the heterotrophic bacteria and fungi, although many other soil animals, described elsewhere in this book, contribute to the breakdown processes.

Simple food chains which begin with a producer organism and consist of three or four links are rarely encountered in isolation. In a soil community one might for instance, pick out a simple chain beginning with a green alga, which is consumed by a collembolan, which is eaten by a predatory nematode. In other words the chain follows the pattern of trophic levels described in the previous paragraph:

green alga   ⟶   collembolan   ⟶   nematode
(primary producer)   (primary consumer)   (secondary consumer)

Inevitably such a simple chain becomes complicated by a number of cross-links. There will be other invertebrate consumers of green algae than collembolans and other predatory animals feeding on collembolans. A further complication of the chain can arise from the fact that few animals are specific in their feeding habits and may use a variety of organisms as food. In addition, there will be parasites attacking all levels of a chain. In these ways a simple chain quickly proliferates into a complicated web with a multiplicity of feeding relationships.

Whether a web is simple or complex, however, the amount of energy available to the rest of the system, is controlled by the primary producers and eventually ends with the decomposers.

Quantifying energy flow

A food chain does not in itself provide a means of measuring the magnitude of the effects of feeding of one species upon another. Macfadyen (1957) points out, that as the links in a chain are built up it soon involves a web consisting of a whole assemblage of organisms and necessitates the separation of those whose absence would alter profoundly the structure of the web from those whose contribution is trivial. Furthermore, in making a food chain quantitative it becomes necessary to estimate the standing crop or quantity of organic matter present at a particular point in time of each population. This is a very poor index of production because metabolic processes cannot be directly related to it nor can the amounts of organic matter passing along a chain be easily labelled and measured.

According to the law of thermodynamics, one form of energy can be transformed into another form of energy and, in the process, no energy is lost from the system and the total amount of energy remains the same. Only in extremely simple systems can a complete energy pathway be identified and the energy budget at each level be measured. This has been attempted in a few ecosystems which are clearly definable and involve a limited number of organisms whose feeding relationships are simple. The work of Odum at Silver Springs, Florida (Odum, 1957) is a classic example.

In complex systems, of which soil is no exception, a complete picture of energy flow would only be possible if data were available on the amount of energy transferred from one trophic level to another and the amount lost at each level by metabolic processes. Accurate information would also be needed on life cycles and food relationships and would involve study over a number of years. It must also be assumed that the system remained reasonably stable.

In any system energy initially enters from the Sun and of the energy arriving in this way about seven-eighths is liberated as heat during the process of photosynthesis, leaving the remaining one-eighth to be converted by plants into substances of high energy content, some being lost by the plants in respiring.

The plant material thus manufactured is made available as food for herbivores or, if it is not consumed, is eventually returned as plant matter which decays and becomes part of the decomposition cycle (Figure 10.1).

The herbivores partly break down their ingested organic matter and in the process liberate energy in the form of heat. In turn the other trophic levels produce heat energy in the same way. The important point is that the energy does not recirculate, as do chemical foods, within a community. The storage of part of the energy intake in the tissues and the liberation of the rest as heat takes place at each trophic level and can be measured at the point at which it passes from one level to another.

Several important points emerge when the dynamic aspects of an ecosystem are considered. The first is that in a stable system the amount of solar energy

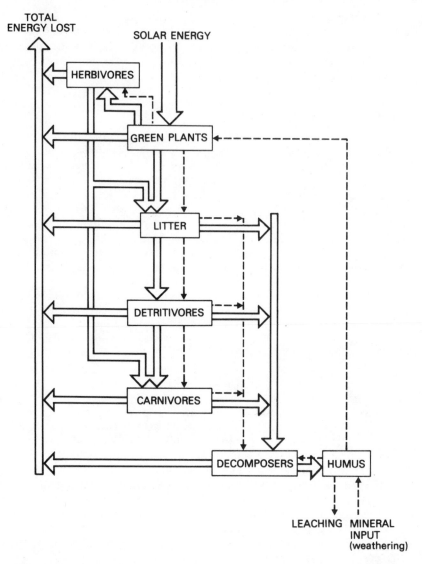

Figure 10.1   Flow of energy (broad arrows) and nutrients (broken arrows) in a greatly simplified soil ecosystem

utilized during photosynthesis is equal to all the energy lost to that system in the respiration of all plants and animals present. Secondly the metabolism of the plants is limited either by solar energy or by nutrients. In other words light intensity and nutrient supplies in the form of inorganic salts are both limiting factors. In tropical forests, plant metabolism proceeds at a far greater rate than in temperate climes where light intensity is less.

Another important generalization is that food organisms breed more rapidly than their consumers, so that a comparatively small stock of food organisms can supply herbivores at a rapid rate. An example of this is to be found in the marine zooplankton. Free organic matter in the sea is largely derived from the waste products of copepods which, having an inefficient metabolic system, excrete up to ten times as much plant food as they actually metabolize. This results in a large amount of nutrient being made available to the phytoplankton through the detritivore chain.

Interesting examples of energy flow paths in various ecosystems are given in Macfadyen (1962b) to which the reader is referred. Macfadyen has also shown that farmers raising bullocks on grassland cannot expect a yield of more than 126 J/m^2 from an initial grass production of over 20 kJ. The bullocks make use of only about one-seventh of the available grass, excreting two-thirds of their food as faeces and a further 90 per cent of that which they assimilate is lost in respiration. The fate of the food not taken by the bullocks depends on how much is consumed by other herbivores and how much will eventually decay and find its way, along with the excreted matter, into the decomposer chain. Some decomposer organisms such as the myriapod, *Glomeris marginata*, produce large quantities of faeces in proportion to their rate of assimilation and their faeces are of great significance biologically, constituting a rich medium for the growth of micro-organisms and also other small arthropods.

This raises the important question of how much, and in what ways, the soil micro- and meso-fauna influence the activities of the micro-flora. Precise figures are not available for all taxonomic groups but it has been found that the addition of small arthropods to soil cultures from which they were previously absent greatly enhances microbial activity by promoting the germination of fungal spores and by the enrichment of the medium by their excreta. In fact they influence microbial activity in a catalytic way by promoting a flow of energy which can be many times that which actually passes through their own bodies.

From Macfadyen's work on the contribution of the micro-fauna to soil metabolism (Macfadyen (1963)), it becomes apparent that the concept of a community consisting of a trophic structure in which herbivores feed on plants and themselves form the food of carnivores, is one of over-simplification. In most food chains, energy flow is greater through the decomposer chain. If the dead plant matter not utilized by herbivores, were allowed to accumulate, the whole community structure would alter because the nutrients contained in that material would not recirculate and would therefore not be available to the living plants.

11 *Man and the Soil*

In previous chapters we have been concerned with the general character of the soil fauna and flora which includes the effects of larger plants on the soil and its inhabitants. Soil ecosystems develop naturally under a wide range of vegetational types such as forests, woodland, and grassland. Man's interference with such natural associations can bring about rapid and drastic changes in the character of the animal and plant communities. These activities can be produced by mechanical or chemical means or by the introduction of species foreign to the community.

Changing the soil

Until the nineteenth century argriculturalists were largely concerned with food production using natural manures assisted by relatively simple mechanical aids.

A classic example of bringing about changes in land almost useless to farming were those wrought on Exmoor by the Knight family during the first part of the nineteenth century. With the Enclosure Act of 1819 the Royal Forest of Exmoor ceased to exist and in that year a Mr John Knight of Worcestershire purchased Exmoor Forest for £50 000. Over 1000 hectares were limed and ploughed up. The 'dry' land was cultivated down to clay subsoil and on 'wet' land the pan was broken by subsoil ploughing. Emulating Coke of Holkham, Knight attempted a four-course rotation of crops, but this was not to succeed on these wind-swept moors. It was Knight's son, Frederick Knight, who tried with success in the more sheltered fields, a policy of root and grass growing to afford fodder for overwintering ewes. Permanent pastures were later achieved by successive liming, burning, and ploughing to produce an annual crop of rape upon which the sheep were folded. These operations, assisted by the trampling of sheep, resulted in the break up of the peaty soil right down to the pan. The subsoil was then ploughed and the land sown with rape and grass seed mixture to form permanent pasture.

This system has become an accepted method for the break-up of peaty moorland soil and the existence on Exmoor of extensive cultivated fields, essential for the overwintering and finishing of sheep stocks for market, is due to the agronomy and perseverance of the Knight family (Figure 11.1).

Modern methods employ highly sophisticated farm machinery, chemical fertilizers, and the raising of large animal stocks which also involve the disposal of vast amounts of slurry. But in these days even farming operations pale before the enormous changes to the soil and landscape that can be wrought almost overnight by the huge bulldozers and earthmovers involved in the construction of major highways with the attendant destruction of natural communities and imposition of new ones by seeding and planting, the final results of which we have yet to experience.

Figure 11.1 Sheep grazing on one of the pasture fields created by the Knight family on Exmoor Forest. Unclaimed forest can be seen in the background

Unwittingly, man can be the cause of serious soil erosion by the felling of trees whose roots bind the soil and create suitable conditions for other plants. By destroying the trees, the whole ecosystem is altered and in more extreme cases, conditions approaching those of a desert can be the result. Areas for instance in Turkey come to mind, where felling of vast forests took place hundreds of years ago. As a result, the water-holding capacity of the soil was decreased followed by a succession of scrub and upland plants. Now the arid stone wastes of today are witness to the ravages of large flocks of goats, practically the only living thing left upon which the present small and scattered population of peasants exist.

Good soil husbandry means supplying the right conditions for plant growth. This involves correct land drainage, the addition to the soil of the minerals needed by the crops to be grown and the production of a good soil structure. All these operations, if correctly applied, will ensure an adequate supply of water and air as well as the elimination of competition by unwanted plants and other organisms harmful to crops.

One of the first direct attempts to eliminate harmful agents was used in the latter part of the nineteenth century to control the root aphis, *Phylloxera*, which attack vines. This was done by the injection of the soil with carbon disulphide – a rather primitive but successful method which resulted in a greatly increased yield of grapes. Heat sterilization of soil in glasshouses is used to control eelworms and other plant parasites. The introduction of chemical fertilizers alters the pH of a soil, while the addition of lime to an acid soil increases the pH. Plants and soil animals each have their own optimum range of pH and any alteration can quickly bring about changes in the soil population which are advantageous for a particular crop but disadvantageous for others. However, since soil organisms are sensitive to soil reaction, effective control is possible of those which are either parasitic or disease-causing by adjusting the soil pH to bring it outside their range of tolerance.

Elimination of pests by chemical means

The pesticide problem is both complex and controversial and in such a brief space it would be impossible to do more than give a few examples and to make a few comments. Before the whole question of pest control, by whatever means, can be adequately assessed, much more information is required since fresh research reveals fresh problems.

The control of pests by the use of chemical compounds may be classified in several ways. For instance the biological effects of herbicides or phytotoxic weed killers, and fungicides are different from those resulting from the use of fumigants and insecticides.

In July, 1976 one of the greatest ecological disasters ever to occur took place in northern Italy. An explosion at a chemical plant scattered the highly toxic substance TCDD (tetrachlorodibenzoparadioxin) over a wide area causing contamination of plants and soil and constituting a major hazard to all life in the locality.

In terms of toxicity, TCDD is in the same league as strychnine, for less than 5 mg is probably lethal to the average human being. Unlike strychnine, however, it takes a considerable time for the effects to be manifested. At present we know little about the long-term effects of TCDD. What we do know is that it is insoluble in water, virtually resistant to biological degradation, and requires a temperature of 800°C to destroy it. Because of its stability, the chemical can remain in the soil, but near the surface, for a long time. Furthermore, TCDD can be translocated by birds and other animals from one area to another. One method of decontamination that has been suggested is to spray a layer of plastic material over the infected soil thus sealing the land against erosion by various agencies. This, however, cannot deal with the problem of translocation which, without constant monitoring, may leave other contaminated areas undetected.

On the whole phytotoxins have little direct effect on the soil fauna although simazine will reduce the population of predatory mites and surface collembolans but have a lesser effect on detritivorous mites, earthworms, enchytraeids, and some insect larvae. Spraying the soil surface or injecting soil with methanal or derris kills all living organisms in the vicinity to a depth of about 15 cm, leaving only those with diapausing stages as survivors. Again, the effects of treatment by insecticides are more variable than either fumigants or phytotoxins on the soil fauna.

Many would-be nature-lovers take the trouble to grow special plants in their gardens to attract bees, butterflies, and other insects. It is strange how few realize that by using toxic sprays to kill nettles, or even by scything them down, they are destroying the food plant of many butterfly larvae such as the peacock, large and small tortoishell, and the red admiral. How unsightly hedgerows and verges become when after spraying, the foliage turns brown! Other indirect sufferers are frogs and toads whose numbers are declining alarmingly in some areas due to their natural foods (flies, slugs, snails, and earthworms) being killed by sprays.

Nowadays stringent efforts are being made to reduce the use of organochlorines such as DDT, aldrin, dieldrin, and heptachlor because of their long-lasting effects in the soil. They can upset the predator–prey relationships of mites and collembolans by reducing the populations of the former with a resultant increase in numbers of collembolans. With the exception of heptachlor, the organochlorins do not seem to affect earthworms, enchytraeids or nematodes. There is, however, the serious and long-term consequence of these pesticides. For instance, although apparently having little effect on earthworms, the worms can accumulate DDT in their bodies. Indeed in the United States in certain areas treated with DDT for Dutch Elm disease, more than 157 ppm have been found in the bodies of worms. Other animals feeding on contaminated worms build up higher levels of the poisons as the food chain progresses.

Observations made on cabbages grown on soil treated with one or other of the organochlorines showed that they were more prone to damage by cabbage root fly than those grown on soil not so treated. The reason appeared to be that there had been a reduction in the numbers of carabid and staphylinid beetles in the first case. Laboratory tests revealed that the beetles and their larvae fed voraciously on the eggs, larvae and pupae of the cabbage root fly. Thus by eliminating the predator without damaging the pest, treatment by these insecticides had actually decreased crop production.

In general, pesticides do little harm to the microbial population of the soil. In fact soil micro-organisms are capable of degrading many pesticides added to the soil, although DDT and dieldrin show considerable resistance and may persist for up to 15 years.

It must not be forgotten that resistant strains can and do arise so that where a pest can withstand a poison, an enormous selection pressure is set up in its favour. Man may contrive to exterminate a pest by chemical means in order to save a crop but the situation rapidly develops into a race between pest and chemist, the latter eventually taking second place.

Biological control of pests

There is no doubt that the best way in which to prevent damage to the soil environment is strictly to limit the use of harmful and long-lasting pesticides. In this contest the wider use of biological methods of controlling pests is greatly to be advocated.

Biological control occurs all around us and is, in fact, another name for the predator–prey relationships which exist within every natural community. The hot, dry summer of 1976 caused recurrent population explosions of aphids in many gardens. A build-up in their numbers was quickly followed by an increase in the incidence of ladybirds and their larvae, the natural predators of aphids. This resulted each time in a dramatic fall in the numbers of aphids.

Like all living organisms, pests can themselves act as hosts to parasites which feed upon them, but to rank as successful, a pest must be able to attack more speedily and efficiently than it is itself attacked.

The term 'biological control' implies the selection of such a parasite to act as a natural enemy in controlling the pest. There are many examples of this type of control. One which has gained notoriety has been used in California where cottony cushion scale, *Icerya puchasi*, was ruining the citrous crop. An Australian ladybird beetle was found to feed on this insect and was introduced with remarkable success. The Australian continent itself has been invaded by many alien plants with their attendant parasites and again biological methods of control have achieved spectacular results. One of the best-known examples used on a commercial scale, was the introduction of large numbers of the Argentinian moth borer. In California the larvae of the moth borer were known to feed on the prickly pear, *Opuntia stricta*. The moth was cultivated and eggs sent out to Australia in large numbers in 1925. Within a few years the prickly pear, which had spread to vast areas of that continent, was brought under control.

Less spectacular have been the efforts to control some of our indigenous crop pests in Britain where we have fewer alien invaders. One example is the introduction of the small chalcid wasp, *Encarsia formosa*, to eliminate the greenhouse whitefly, *Trialeurodes vaporariorum*, which attacks tomatoes and cucumbers (Figure 11.2). In this case both host and parasite are alien species. Control of insect pests can sometimes be achieved by cultivating certain plants

(a)　**Figure 11.2**　(a) Glasshouse whitefly, *Trialeurodes raporariorum* and (b) scale bug. *Encarsia formosa*, an effective parasite of *T. vaporariorum*. (*Photo courtesy of the Glasshouse Crops Research Institute.*)

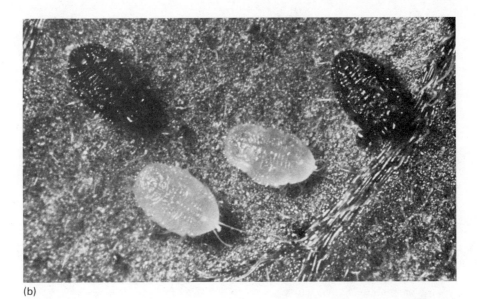

(b)

repellent to the pest. In the example just quoted, the numbers of *T. vaporariorum* can be greatly reduced in glasshouses by interplanting shoofly plants, *Nicandra physaloides*, with the host plants.

Australia has another problem, that of disposing of cattle-dung pads. In the absence of dung beetles, these pads can remain on pastures for up to five years, the rich grass growing round the pads being left untouched by cattle. Australia has no native dung beetles but recently an attempt has been made to introduce species from the African continent to bring about the dispersal of the pads.

Mention must also be made of the more recent success in controlling the red spider mite, *Tetranychus urticae*, by another mite, *Phytoseiulus persimilis* (Figure 11.3). *T. urticae* can undergo ten or more generations (Figure 11.4) in a year under glasshouse conditions, passing through the stages

$$egg \rightarrow larva \rightarrow protonymph \rightarrow deutonymph \rightarrow adult.$$

The predator goes through the same stages but twice as fast, feeding in all its stages (except as eggs) on *T. urticae*, even attacking its eggs. Furthermore, *T. urticae* overwinters as a diapause stage in the soil. Diapause is induced by a reduction in light intensity and since all diapausing mites of the species are female, light may have a genetic influence on the eggs causing a change of sex. The predator mite has no diapause stage but will maintain a low population in winter so long as its prey is available. One interesting fact about red spider mite is that it occurred, although not in pest proportions, on outdoor fruit crops. The introduction of toxic sprays such as DDT to control other pests and to which the red spider mite is immune, eliminated the slow-breeding insect predators of the mite with a resultant increase in its numbers.

Some of the worst soil pests are eelworms of many kinds. The inoculation of soil with spores of predatory fungi belonging to the Hyphomycetes, has been used successfully to kill the root knot eelworm which parasitizes begonias and also against *Strongyloides papillosus*, a gut nematode of sheep. The fungus

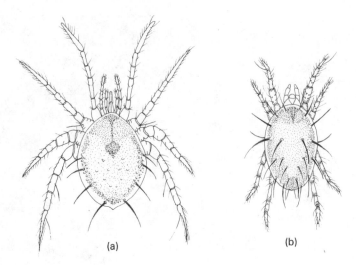

(a)   (b)

Figure 11.3 (a) Red spider mite, *Tetranychus urticae*, 0.5 mm, a serious pest of greenhouse crops and (b) its predator, *Phytoseiulus persimilis*, 0.75 mm

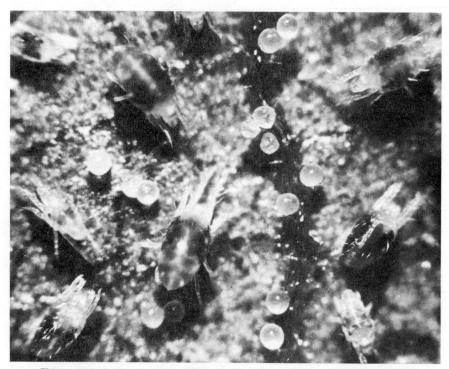

Figure 11.4 Red spider mite, *Tetranychus urticae*. The photograph shows a female in the centre, a male lower right and spherical eggs. (*Photo courtesy of the Glasshouse Crops Research Institute.*)

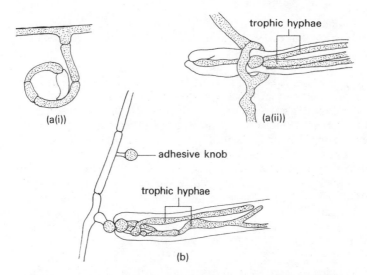

Figure 11.5 Two species of hyphomycete: (a) *Dactylaria candida*. The mycelium forms numerous non-constricting rings (i). Once a nematode attempts to pass through the ring it becomes wedged and cannot withdraw (ii). It is finally killed by the invading trophic hyphae which penetrate the body of the eelworm, gradually absorbing its substance and eventually leaving an empty skin. (b) *Dactyllela ellipsospora*. The mycelium forms a number of adhesive knobs to which eelworms stick. Invading hyphae act in the same way as in *D. candida* to kill the eelworm

forms a constrictive ring around the eelworm which is mortal (Figure 11.5).

The dramatic successes achieved in the control of the prickly pear in Australia and the cottony cushion scale in California may have led to the assumption that methods of biological control are the answer to most pest problems. However, this is far from the case since success depends upon finding a natural enemy in sufficient numbers and sufficiently powerful to overcome the pest. But the occurrence of such predators is small. Probably the solution lies in making use of both chemical and biological methods, the one supplementing the other.

Pollution of the soil

Whereas the use of pesticides is a deliberate attempt to control those organisms considered by man to be harmful to his crops, pollution in all its forms arises as a direct result of an increase in human populations. Perhaps Lassa fever or some other virus disease will prove effective in controlling *our* numbers! Whichever way the problem is regarded, the effects of pollution on the countryside become obvious where there are large towns or industrial activity with all the attendant problems of sewage and waste disposal.

Examples of air pollution and water pollution are legion. Both have only an indirect but nevertheless important effect on the soil and should be mentioned. Much has been written about air pollution and the incidence of 'smog' in industrial cities. Atmospheric pollution involving a rise above a certain level of sulphur dioxide, a reduction in light intensity or deposits of soot on leaves and bark, can be deleterious to plant growth. Some of these pollutants can be washed into the soil, although amounts entering in solution are probably slight.

Exhaust gases from motor vehicles contain high concentrations of carbon monoxide and lead. The latter has been found to accumulate in vegetation and soil along road verges. Although little work has been done on the possibility of lead entering food chains, in the United States earthworms collected near main roads have been found to contain toxic quantities of lead and other metals in concentrations that would be fatal to young pigs, although the worms themselves appear to be unaffected.

Refuse dumps, usually sited on hillsides, can be another source of soil pollution lower down the slopes. Poisonous chemicals such as compounds of sodium, phosphorus, and arsenic find their way into the soil from industrial waste products. Farm fertilizers are another hazard for, washed down into the soil, they produce eutrophication of ponds and streams by altering the balance of salts thereby causing a rapid growth of algae.

Our responsibilities

Along cliff paths, in national parks and other beauty spots, the sheer effects of trampling by human feet can cause severe erosion (Figure 11.6). In fact human pressures on the countryside, of many kinds, are increasing every year. We must be thankful for the various organizations which take a responsible attitude to the problems of conservation and to the management of the land. As individuals, it should be our business to inform those who are not yet aware, that they too are users of the land and must take their share of responsibility.

Figure 11.6   A cliff path showing the effects of trampling and erosion of the cliff. (*Photo by courtesy of Peter Beale.*)

Glossary of terms not explained in the text

aedeagus – organ possessed by some male arthropods for the introduction of spermatozoa into the female

aestivation – dormancy during dry season

anabiosis – inert condition assumed during unfavourable conditions

autotrophic – (of organisms) capable of manufacturing organic material from inorganic substances (cf. heterotrophic)

campodeiform – type of larva with flattened body and bearing tail filaments

cephalothorax – fore part of the body of an arachnid corresponding to the head and thorax of an insect

cerci – paired appendages, sometimes sensory, at the hind end of some insects

chelicerae – jaws of an arachnid

chemoreceptor – a chemical sense organ capable of perceiving taste or smell

chitinized – presence of chitin, a horny skeletal material produced by arthropods, in the covering layer of the body

cilium – a cytoplasmic thread, capable of movement, projecting from the surface of a cell

cryptozoic – (of animals) inhabiting crevices under stones or in soil

diapause – state of dormancy with arrested growth and low rate of metabolism

ectotrophic – (of mycorrhizas) fungus mycelium growing outside the cells of the root cortex (cf. endotrophic)

endotrophic – (of mycorrhizas) fungus mycelium growing within the cells of the root cortex (cf. ectotrophic)

epicuticle – superficial layer of the cuticle commonly containing a waxy substance for waterproofing

exuvia – structural material shed by an organism when it moults

fossorial – burrowing, used in burrowing

geophiles – organisms inhabiting the soil

haltere – small knob or projection representing a hind wing in two-winged flies (Diptera)

heterotrophic – (of organisms) incapable of synthesizing organic food from inorganic substances (cf. autotrophic)

hypertrophy – an increase in size of a tissue or organ

hypopus – resistant, dispersal stage of some mites

instar – larval, nymphal or other developmental stage of an arthropod

mycophagous – fungus-eating

ocellus – light perceptor or simple type of eye incapable of true vision

parthenogenesis – development of unfertilized eggs

pedipalp – leg-like tactile organ situated in front of the legs in arachnids

phoresy – utilization of an animal by another for transport

photonegative response – a movement away from light

phytophages – animals which feed on plant material

phytotoxin – a substance poisonous to plants

receptaculum – organ for the reception of spermatozoa

saprophages – organisms which feed on carrion

sclerotized – hardened, toughened

spermatophore – bag containing spermatozoa, secreted by the male in some arthropods

tergum – dorsal plate of a body segment of arthropods

Bibliography

Anderson, J. M. and Macfadyen, A. (eds) (1976). The role of terrestrial and aquatic organisms in decomposition processes. *Brit. Ecol. Soc. Symposium No. 17.* Blackwell Scientific.

Bechyne, J. (1956). *Guide to beetles.* Thames and Hudson.

Blower, J. G. (1955). Millipedes and centipedes as soil animals, In: *Soil Zoology.* D. K. McE. Kevan (ed.). 138–51.

Bornebusch, C. H. (1930). The fauna of forest soil, Forst. *Fors Vaes Danm.* II.

Bristowe, W. H. (1968). *The world of spiders.* Collins New Naturalist.

Burges, A. (1958). *Micro-organisms in the soil.* Hutchinson Univ. Library.

Burges, A. and Nicholas, D. P. (1961). Use of soil sections in studying amounts of fungal hyphae in soil. *Soil Sci.* **92**, 25–9.

Carson, R. (1961). *Silent spring.* Hamish Hamilton.

Chu, H. F. (1949). *How to know the immature insects.* Wm. C. Brown, Iowa.

Cholodney, N. C. (1930). Über eine neue Methode zur Untersuchung der Boden-mikroflora. *Arch. Mikrobiol.* **1**, 620–52.

Cloudsley-Thompson, J. L. (1952). Studies in diurnal rhythms. II. Changes in the physiological responses of the woodlouse *Oniscus asellus* to environmental stimuli. *J. exp. Biol.* **29**, 295–303.

Cloudsley-Thompson, J. L. (1960). *Animal behaviour.* Oliver and Boyd.

Cloudsley-Thompson, J. L. and Sankey, J. (1961). *Land invertebrates.* Methuen.

Cloudsley-Thompson, J. L. (1967). *Micro-ecology.* Inst. of Biol. Studies in Biology, No. 6. Arnold.

Cloudsley-Thompson, J. L. (1968). *Spiders, scorpions, centipedes and mites.* Pergamon Press.

Darlington, A. and Leadley Brown, A. (1975). *One approach to ecology.* Longman.

Darwin, C. (1881). *The formation of vegetable mould through the action of worms, with observations of their habits.* Murray.

Edney, E. B. (1951). The evaporation of water from woodlice and the millipede *Glomeris. J. exp. Biol.* **28**, 91–115.

Edney, E. B. (1954). Woodlice and the land habitat. *Biol. Rev.* **29**, 185–219.

Edney, E. B. (1968). Transition from water to land in Isopod crustaceans. *Am. Zool.* **8**, 309–26.

Edwards, C. A. and Fletcher, K. E. (1971). A comparison of extraction methods for terrestrial arthropods. In: *Methods of study in quantitative soil ecology: population, production and energy flow.* J. Phillipson, (ed.). 150–85.

Edwards, C. A. and Lofty, J. R. (1972). *Biology of earthworms.* Chapman and Hall.

Evans, A. C. and Guild, W. J. McL. (1947). Studies on the relationship between earthworms and soil fertility. 1. Biological studies in the field. *Ann. appl. Biol.* **34**, 307–30.

Evans, G. O. (1955) Identification of terrestrial mites. In: *Soil Ecology.* D. K. McE. Kevan (ed.). 550–61. Butterworths.

Evans, C. O., Sheals, J. C. and MacFarlane, D. (1961). *Terrestrial Acari of the British Isles. 1. Introduction and biology.* B. M. Nat. Hist.

Fletcher, W. W. (1974). *The pest war.* Blackwell.

Gray, T. R. G. and Williams, S. T. (1971). *Soil micro-organisms.* Oliver and Boyd.

Griffin, D. N. (1963) Soil moisture and the ecology of soil fungi. *Biol. Rev.* **38**, 141–66.

Guild, W. J. Mc. L. (1951). The distribution and population density of earthworms (Lumbricidae) in Scottish pasture fields. *J. Anim. Ecol.* **20** (1), 88–97.

Guild, W. J. Mc. L. (1955). Earthworms and soil structure. In: *Soil Ecology*. D. K. Mc.E. Kevan (ed.). 83–98. Butterworths.

Harley, J. L. and Brierley, J. K. (1963). A method of estimating oxygen and carbon dioxide concentration in the litter layer of beech woods. *J. Ecol.* **41**, 385–7.

Hurley, D. E. (1968). Transition from water to land in amphipod crustaceans. *Am. Zool.* **8**, 327–53.

Jackson, R. N. and Raw, F. (1966). *Life in the soil*. Inst. of Biol. Studies in Biology No. 2. Arnold.

Kevan, D. K. McE. (ed.) (1955). *Soil Zoology*. Butterworths.

Kevan, D. K. McE. (1962). *Soil animals*. Witherby.

Lakhani, K. H. and Satchell, J. E. (1970). Production by *Lumbricus terrestris* (L). *J. Anim. Ecol.* **39**, 473–92.

Lee, D. L. and Atkinson, H. J. (1976). *Physiology of nematodes* (2nd edn). Macmillan.

Macfadyen, A. (1961). Improved funnel-type extractors for soil arthropods. *J. Anim. Ecol.* **30**, 171–84.

Macfadyen, A. (1962). Soil arthropod sampling. In: *Advances in ecological research*. J. B. Cragg (ed.). **1**, 1–34.

Macfadyen, A. (1962). Energy flow in ecosystems. In: *Grazing in terrestrial and marine environments*. D. J. Crisp (ed.). Blackwell.

Macfadyen, A. (1963). The contribution of the microfauna to total soil metabolism. In: *Soil organisms*, J. Doekson and J. van der Drift (eds). North-Holland Pub. Co.

Macfadyen, A. (1963). *Animal ecology: Aims and methods* (2nd edn). Pitman.

Macfadyen, A. (1964). Energy flow in ecosystems and its exploitation by grazing. In: *Symposium on grazing*. D. J. Crisp (ed.). British Ecol. Soc. Blackwell Scientific.

Murphy, P. W. (1962). *Progress in soil zoology*. Butterworths.

Nef, L. (1957). État actuel des connaissances sur le rôle des animaux dans le decomposition des litières de fôrets. *Agricultura (2)*, **5**, 245–316.

Nielson, C. O. (1953). Studies in Enchytraeidae. A technique for extracting Enchytraeidae from soil samples, *Oikos* **4**, 187–96.

O'Connor, F. B. (1967). The Enchytraeidae. In: *Soil Biology*. A. Burges and F. Raw (eds). 212–57. Academic Press.

Odum, H. T. (1957). Trophic structure and productivity of Silver Springs, Florida. *Ecol. Monogr.* **27**, 55–112.

Peters and van Slyke, D. D. (1956). *Quantitative clinical chemistry methods. Vol. 2*. Williams and Wilkins, Baltimore.

Phillipson, J. (1966). *Ecological energetics*. Inst. of Biol. Studies in Biology. No. 1. Arnold.

Phillipson, J. (ed.) (1971). *Methods of study in quantitative soil ecology: Population, production and energy flow*. Blackwell Scientific.

Raw, F. (1959). Estimating earthworm populations by using formalin. *Nature, Lond.* **184**, 1661–2.

Raw, F. (1962). Studies of earthworm populations in orchards. 1. Leaf burial in apple orchards. *Ann. appl. Biol.* **50**, 389–404.

Russell, E. J. (1957). *The world of the soil*. Collins New Naturalist.

Russell, E. J. (1961). *Soil conditions and plant growth*. Rewritten by E. W. Russell. Longman.

Satchell, J. E. (1955). Some aspects of earthworm ecology. In: *Soil Zoology*. D. K. McE. Kevan (ed.). 180–201. Butterworths.

Satchell, J. E. (1971). Earthworms. In: *Methods of studying quantitative soil ecology: Population, production and energy flow*. J. Phillipson (ed.). Blackwell Scientific.

Savory, T. H. (1955). *The world of small animals*. U.L.P.

Singh, B. N. (1946). A method of estimating the numbers of soil Protozoa especially amoebae, based on their differential feeding on bacteria. *Ann. appl. Biol.* **33**, 112–19.

Skoczen, S. (1958). Tunnel-digging by the mole, *Talpa europea. Acta Theriologica, Bialowieza,* **2**.

Southwood, T. R. F. (1966). *Ecological methods.* Methuen.

Stamp, L. D. (1955). *Man and the land.* Collins New Naturalist.

Stamp, L. D. (1962). *The land of Britain: Its use and misuse.* Longman.

Townsend, W. N. (1973). *An introduction to the scientific study of the soil.* Arnold.

Wallwork, J. A. (1970). *Ecology of soil animals.* McGraw-Hill.

Webley, D. (1964). Slug activity in relation to weather. *Ann. appl. Biol.* **53**, 407–14.

For identification keys

Chu, H. F. (1949). *How to know the immature insects.* Wm. C. Brown. Iowa.
    (for keys to larval coleoptera and diptera)

Cloudsley-Thompson, J. L. and Sankey, J. (1961). *Land invertebrates.* Methuen.
    (for key to free-living nematodes)

Eason, E. H. (1964). *Centipedes of the British Isles.* Warne.

Edwards, C. A. and Lofty, J. R. (1967). *Biology of earthworms.* Chapman and Hall.
    (for key to terrestrial earthworms)

Janus, N. (1965). *Land and freshwater molluscs.* Burke.
    (for key to snails)

Quick, H. E. (1960). British slugs. *Bull. Br. Mus. Nat. Hist.,* **6**, 103–226.

Index

Numbers in bold type refer to the figures.

Acari (mites), general, 46–9, **5.3**
  key to, 33–4
  seasonal fluctuations, **5.4**
Actinomycetes, 36, **4.2**
  estimation of numbers, 78
Agriolimax reticularis, 68
Agriotes sp. as crop pests, 56
  as soil tunnellers, 74
Algae, general, 39–40
Allelobophora sp., 64, 67
Ammophila sabilosa, habits and burrow
  construction, 73
Amphipoda, 60
Anamorphosis, in chilopod development, 63
Annelida, 64–8
Ant, 54, 57
Ant lion (*see Myrmeleon* sp.)
Aphids, 58, biological control of, 104–5,
  11.2(a), (b)
Arachnida, 44–5
Arcella, 40
Arion hortensis, 68
Armadillidium vulgare, 58, 60, **6.3(d)**
Armallaria mellea, 38
Arthropods, key to, 25–30
Auger, 15
Azotobacter, and N-cycle, 88
  estimation of numbers, 78

Bacteria, 35–6, **4.1**
  estimation of numbers by staining, 77, by
    dilution plate method, 78
Baermann funnel, 81–2, **8.3**
Beetles (*see* Coleoptera)
Berlese–Tullgren funnel, 80, **8.2**
Bibio sp., larvae as coprophagous feeders,
  56, **6.2(c)**
  as soil tunnellers, 74
Bimastos eiseni, 67
Biological control of pests, 103–7
Blow fly (*see Calliphora* sp.)

Calliphora sp., larvae as carrion feeders, 56
Calluna vulgaris, mycorrhizal association, 86,
  9.1
Capillary water in soil, 9
Carabidae, 54
Carbon, content of in soil, 4
Carbon cycle, 86–8, **9.2**
Centipedes (*see* Chilopoda)
Chaetonotus sp. 52, **5.7(b)**
Chalcid wasp (*see Encarsia formosa*)
Cheilobus quadrilabitus, phoretic larvae, 41

Chelonethi, courtship and life history, 45, **5.2**
  feeding, 76
Chemical control of pests, 102–3
Chilopoda, 61–3, **6.7**
Cicadas, 58, nymphal adaptations and
  construction of earthen chimneys, 72,
  7.3(c), (d)
Cicindela campestris, larva of 54, **6.1(a)**
  soil excavation, 70
Clay, content in soil, 7
  pore space, 8
  moisture content of, 8
  drying boundary curve of, 10, **1.4**
Click beetle (*see Agriotes* sp.)
Clostridium pasteurianum, factors affecting
  distribution, 12
Cockchafer beetle (*see Melalontha
  melalontha*)
Coleoptera, 54–6, **6.1**
  key to larvae of, 31–3
Collembola, adaptations, 74, **7.5**
  as agents of humification, 4, 91
  distribution, 50
  general, 47–51, **5.5**
  regulation of numbers, 50
Colpidium, **4.5(b)**
Cottony cushion scale (*see Icerya puchasi*)
Cranefly (*see* Tipulidae)
Crustacea, general, 58

Dactylaria candida, control of eelworms,
  11.5(a)
Dactyllela ellipsospora, control of eelworms,
  11.5(b)
Darwin, Charles, and earthworm casts, 67
Decticus verrucivorus, and burrow
  excavation, 72, **7.3(b)**
Dendrobaena sp. 64, 67
Dermaptera, 56, 58
Difflugia, 40, **4.5(a)**
Dilution plate method of estimating
  micro-organisms, 78
Diplopoda, 61, 63–4
  and humification, 64, 91
  as tunnellers, 74
Diptera, general, 56
  key to larvae, 31–3
Dor beetles (*see* Scarabidae)
Dung beetles (*see* Scarabidae)
Dysdera crocata, woodlice as prey, 44, **5.1**

Earthworms, as tunnellers, 74
  and Charles Darwin, 67

Earthworms *cont.*
  burrows, 67–8, **6.11**
  distribution, 67, **6.10**
  effect of moisture and temperature on, 64–5
  extraction of, 83–4
  predators of, 68
  reproduction, 64
  soil types and populations, 66
  tolerance of different species to pH, 64, **6.9**
Earwigs (*see* Dermaptera)
Eimer's organs in *Talpa europea*, 71
Eisenella sp. 64
Encarsia formosa, as biological control, 104, **11.2(b)**
Enchytraeidae, general, 68
  extraction of, 79, **8.1**
Energy, production of by a community, 96
  quantifying flow of, 97–9, **10.1**
Epimorphosis, in chilopod development, 63
Euglena viridis, 40, **4.5(c)**
Extraction, methods of, 79–83
  by dry and wet funnels, 80–2, **8.2, 8.3**
  by flotation, 80
  by Nielson extractor, 82, **8.4**
  by wet-sieving, 79, **8.1**
  comparison of methods, 83
  in the field, 83–4

Fabre, J. H., and observation of *Ammophila*, 73
Fannia sp., larvae as coprophagous feeders, 56, **6.2(d)**
False scorpions (*see* Chelonethi)
Fauna of soil, classification of, 21–3, **3.1**
Feeding mechanisms, 75
Field capacity, definition of, 8
Filth fly (*see Fannia* sp.)
Flies (*see* Diptera)
Food chains, webs, 96–7
Formica rufa, nests of, 57
Formica sanguinea, 54
Fungi, general, 37–9
  and mor soils, 5
  estimation of numbers, 77–9
  symbiotic association with roots, 85
Fungus gnats (*see* Mycetophilidae)

Garden snail (*see Helix aspersa*)
Gastrotricha, 52, **5.7(a)**
Geotrupes sp. 54–6, **6.1(d)**
  adaptations for soil excavation, 70, **7.1(a), (b)**
Glomeris marginata, 64, **6.8(b), (c)**
  and faecal production, 99
Gravitational water in soil, 9
Great green bush cricket (*see Tettigonia viridissima*)
Ground beetles (*see* Carabidae)
Gryllotalpa gryllotalpa, 58
  fossorial adaptations, 71–2, **7.2(b)**

Habrotrocha tridens excedens, **5.6**
Hairy back (*see* Gastrotricha)
Helix aspersa, 68
  estimation of populations, 84
Heterodera rostochiensis, 43, **4.8**

Humus, formation of, 4, 90–1, **9.4**
  estimation of, 16
Hygroscopic water, 9
Hymenoptera, general, 56–7
Hypogastrura sp., 74, **7.5(b)**

Icerya puchasi, biological control of, 103–4
Insects, general, 54–8
Isopoda, effect of temperature on rate of transpiration, 58, **6.4**
  response to light, 58
  water loss in, 58
Iulid millipedes, 64, **6.8(a)**

Keys to soil animals, 23–34
  to 'worms', 24–5
  to arthropods, 25–30
  to beetle and fly larvae, 31–3
  to mites (Acari), 33–4
Knight, John and reclamation of Exmoor Forest, 100, **11.1**

Lava, colonization of, 1, **1.1**
Lithobius forficatus, habits, 61 **6.7(a)**
  life history, 63
Loam, drying boundary curve of, 10, **1.4**
  moisture content of, 8
Lomenchusa, larvae in nests of *Formica sanguinea*, 54
Lucanus cervus, larva of, 56
Lumbricus terrestris, distribution, 67, **6.10**
  and leaf burial, 65
Lyristes plebejus, **7.3(c)**

Macfadyan's controlled draft funnel, 80–1, **8.2**
Macrobiotus sp., 52, **5.7(a)**
March fly (*see Bibio* sp.)
Melalontha melalontha, larva of, 56, **6.1(e)**
Micro-organisms, association with roots, 85
  effect of temperature gradients on, 13
  and capillary water, 10
  and C-cycle, 88
  and gaseous content of soil, 12
  distribution, 43
  estimation of numbers, 77–9
  and humification, 4–5, 90–1
  and suction pressure, 10
Milax budapestensis, 68
Millipedes (*see* Diplopoda)
Minerals, estimation of in soil samples, 16–17
Mites, as agents of humification, 4
  nutrition, 76
  vertical migration, 13
Mole (*see Talpa europea*)
Mole cricket (*see Gryllotalpa gryllotalpa*)
Molluscs, 68–9, cellulose digestion and crumb formation, 69
  estimation of populations, 84
Mor soils, 5–6, **13(a)**
Mothflies (*see* Psychoda)
Mucor sp., 37
Mull soils, 5–6, **1.3(b)**
Mycetobacterium tuberculosis, 37
Mycetophilidae, larvae feeding on fungal mycelia, 56, **6.2(b)**

Mycorrhiza, 85–6
and *Calluna*, 86
Myriapoda, general, 61–4
and water relations, 75
Myrmeleon sp., excavation of pit, 73,
7.4(a), (b)
feeding, 75–6

Necrophloeophagus longicornis, habits and
water loss, 61, **6.7(b)**
as soil tunneller, 74
Necrophorus vespillo, 54
Nematodes, 41–3
adaptations to drought, 75
as vectors of disease, 41
biological control of, 105, 106
regulation of numbers, 43
structure of buccal capsule, 41, **4.7**
Nicandra physeloides, as biological control,
105
Nielson extractor, 82, **8.4**
Nitrobacter agilis, and N-cycle, 88
Nitrogen content of soil, 4
Nitrogen cycle, 88–9, **9.3**
Nitrosomonas europea and N-cycle, 88

Oniscus asellus, 58, **6.3(b)**
effect of temperature on rate of
transpiration, **6.4**
Opuntia stricta, biological control of, 104
Orchesella sp., 74, **7.5(c)**
Organochlorines, 103
Osmotic water in soil, 9

Pauropoda, 61
Pauropus sp., **6.6(a)**
Penicillium sp., 39, **4.4**
Pests, control of, 101–7
biological control of, 103–7
chemical control of, 102–3
Philoscia muscorum, 58, **6.3(c)**
effect of temperature on rate of
transpiration, **6.4**
Phosphorus cycle, 88
Phragmosis in larva of *Cicindela campestris*,
54
Phylloxera, control of in vines, 101
Phytoseiulus persimilis, as biological control,
105, **11.3(b)**
Phytotoxins, use of, 102
Pill millipede (*see Glomeris marginata*)
Podzols, 5, **1.3(a)**
profile, **1.5**
Pollution of soil, 107–8
Polydesmus sp. 64, **6.8(d)**
Populations, estimation of, 77–84
metabolism, 94, 96
regulation of, 92
sampling in the field, 83–4
Porcellio scaber, 59, **6.3(a)**
effect of temperature on rate of
transpiration, **6.4**
Potworms (*see* Enchytriaeidae)
Prickly pear (*see Opuntia stricta*)
Protozoa, general, 40–1
adaptations to drought, 75
estimation of populations, 78

Pseudomonas and N-cycle, 88, 89
Psychoda sp., larvae as coprophagous
feeders, 56
Pythium debaryanum, life history of, 37,
4.3

Red spider mite (*see Tetranychus urticae*)
Relative humidity of soil atmosphere,
definition of, 10
Rhabditis sp. 41, **4.6**
Rhizosphere, 85
Roots, environment of, 85–6
Rossi–Cholodney slide technique, 77–8
Rotifera, 51–2, **5.6**
adaptations to drought, 75
Rove beetles (*see* Staphylinidae)

Sand, content in soil, 7
moisture content of, 8
drying boundary curve of, 10, **1.4**
Sand wasp (*see Ammophila sabilosa*)
Scarabaeidae, 54, as soil excavators, 70, **7.1**
Scatopsidae, larvae as coprophagous feeders,
56
Scutigera coleoptrata, 61–3, **6.7(c)**
Scutigerella immaculata, 61, **6.6(b)**
Sedimentary rock, 2, **1.2**
Sense organs, and adaptation to subterranean
life, 75
Shoo-fly (*see Nicandra physaloides*)
Silt, content in soil, 7
Slugs (*see* Molluscs)
Soil, formation of, 2–4
adaptation to life in, 70–6
C:N ratio, 4
conditions for life in, 7–13
crumb structure, 7, 8, and mollusca, 69
decomposition of organic content, 4
excavators, 70–3, **7.1, 7.2, 7.3, 7.4**
horizons, 5
pH of, 15–16
profiles, 4, 5, making of, 14
texture, 7
tunnellers, 74
Soil atmosphere, 10–12
and effect of temperature, 12
composition of, 11
measurement of gases, 17–18, **2.2**
relative humidity of, 10
Soil corer, 14, **2.1**
Soil temperature, 12–13
measurement of, 18–20, **2.3, 2.4, 2.5, 2.6**
Soil water, 8–11
adaptations of fauna to, 75
and field capacity, 8
and waterlogging, 8
estimation of, 16
loss by evaporation, 9
Solitary bees, underground nests of, 56
Stag beetle (*see Lucanus cervus*)
Standing crop, definition of, 94
Staphylinidae, 54, **6.1(b), (c)**
Streptomyces sp., **4.2**, *S. griseus*, 37, *S.*
aureofaciens, 37, *S. scabies*, 37
Suction pressure (pF) in soil, 9–10
Sulphur cycle, 89
Symphyla, 61

Tachypodiulus niger, **6.8(a)**
Talitroides dorrieni, **6.5**, occurrence in Great
    Britain, 60
  diet, 60
  evolution from supralittoral species, 60
Talpa europea, fossorial adaptations, 70–1,
    7.2(a)
Tardigrada, 52
  adaptations to drought, 75
Testacella sp., predation, 68
  T. haliotoidea, **6.12**
Tetranichus urticae, biological control of,
    105, **11.3(a)**, 11.4
Tettigonia viridissima, and burrow
    excavation, 72, **7.3(a)**
Thermistor, 18, **2.3, 2.5**
Tiger beetle (*see Cicindela campestris*)
Tipulidae, larvae of, 56, **6.2(a)**
  as soil tunnellers, 74

Trialeurodes vaporariorum, biological control
    of, 104–5, **11.2(a)**
Tullbergia sp. 74, **7.5(a)**

Viruses, 35

Wart-biter (*see Decticus verrucivorus*)
Water bears (*see* Tardigrada)
Wet-sieving, 79–80
Wheatstone bridge, 18, **2.4**
Whitefly (*see Trialeurodes vaporariorum*)
Whiteworms (*see* Enchytraeidae)
Wilting point, definition of, 8
  and evaporation of soil water, 9
Wireworm (*see Agriotes*)
Wood ant (*see Formica rufa*)
Woodlice (*see* Isopoda)

Xiphinema sp. and virus transmission, 35

KING ALFRED'S COLLEGE
LIBRARY